What Freud Didn't Know

What Freud Didn't Know

A Three-Step Practice
for Emotional Well-being
through Neuroscience
and Psychology

TIMOTHY B. STOKES, PH.D.

Rutgers University Press

New Brunswick, New Jersey, and London

LIBRARY OF CONGRESS CATALOGING-IN-PUBLICATION DATA

Stokes, Timothy B., 1948–
What Freud didn't know : a three-step practice for emotional well-being through neuroscience and
   psychology / Timothy B. Stokes.
      p.   cm.
   Includes bibliographical references and index.
   ISBN 978-0-8135-4640-7 (hardcover : alk. paper)
   1. Neuropsychiatry.   2. Amygdaloid body.   3. Affective neuroscience.   I. Title.
   [DNLM: 1. Amygdala—physiology.   2. Emotions—physiology   3. Neuropsychology—
      methods.   4. Psychotherapy—methods.   WL 314 S874w 2009]
   RC341.S87 2009
   616.89—dc22   2009006060

A British Cataloging-in-Publication record for this book is available from the British Library.

Visit our Web site: http://rutgerspress.rutgers.edu

Manufactured in the United States of America

Dedicated to Wendy, that most delightful collaborator in my life, enriching countless endeavors. Her editing, support, and good-humored skepticism made this book better by far than it would otherwise have been.

CONTENTS

___TABLES

PREFACE

During the thirty-plus years that I have been practicing psychotherapy, a humbling amount of clinical and neurobiological research has demonstrated new and refined methods for overcoming psychological problems. This book is rooted in that research. In the eyes of most readers, this book will appear to be a major departure from the traditional views of psychotherapy and psychological change strategies. Rather than thinking of what is presented here as a new breakaway therapy, however, it is more fitting to think of this book as describing a tool born out of a brew of important mainstream research.

The concept of amygdala scripts owes its existence to my struggle with a theoretical conundrum that existed in the psychotherapy world as far back as the 1970s, when I first started out as a graduate student and psychotherapist. At that time, neurobiology had not yet found a way to explore the processes associated with psychological change, and "mindfulness" was a term that most psychologists had never heard. My work and that of most other psychotherapists was strongly influenced by the "insight-oriented" therapies of Sigmund Freud. Those of us who were apprentice psychotherapists were bolstered by research that demonstrated the effectiveness of such therapies. However, in the 1970s there was also fierce competition between two scientifically valid but seemingly contradictory ways of understanding the underpinnings of the process of psychological change. One school of thought championed the insight therapies inspired by Freud, and the other championed the cognitive therapies that were being formulated in the books and research of Albert Ellis and Aaron Beck. According to the insight-oriented school, the unconscious plays a primary role in the creation of psychological problems and a cathartic release of emotions is central to psychological healing. This view stands in sharp contrast to cognitive therapy, which was an outgrowth of the behavioral school of psychology—a school that found little use for the concept of the unconscious, nor did it value the cathartic release of emotions as therapeutic. The cognitive-behavioral school of psychotherapy perceived psychological problems as the straightforward result of learned responses.

The views held by the cognitive behaviorists seemed so fundamentally different from those held by the "insight therapists" that it was hard to imagine how common ground could exist between the two. There was much debate as to which school of therapy would ultimately prove itself as the preferred treatment: those that promoted emotional catharsis and insight as central mechanisms for positive psychological change, or those that favored hard behavioral principles largely derived from experiments with animals. The result was a tension felt by myself and many other psychotherapists—a tension that eventually set the stage for the development of this book.

In my undergraduate studies, I had minored in chemistry and had garnered a very healthy respect for strictly empirical research— a background that inevitably drew me to appreciate the behaviorists. But I had left the rigor of the physical sciences in favor of a fascination with philosophy and with the richness of human experience. I felt that behavioral psychology was hobbled by restricting itself to appreciating only those aspects of the human mind that could be accessed through what I perceived as the accurate but myopic reach of the research laboratory. I had developed a more far-ranging interest in the human mind in all of its manifestations. I didn't want my psychotherapy practice to be limited to behavioral learning principles, especially when other approaches were proving effective. And so I chose to focus my graduate studies on the practices of insight therapy.

However, fortune saved me from becoming too one-sided—I was introduced to Dr. Jenny Steinmetz. Jenny was a very astute behavioral therapist, a writer, and also a faculty member at my graduate school. I was drawn to her quiet intelligence and her amazing ability to be forthright while avoiding entanglement in school politics (a predominant feature of my, and most, graduate schools). She became a mentor to me—a cognitive-behaviorist mentor—who was undaunted by my strong insight-therapy leanings. Through countless discussions and much debate, Jenny succeeded in impressing upon me a great deal about the human application of behavioral principles. By the time I graduated, Jenny had planted theoretical seeds that I could not completely ignore, as I continued to learn more about what constitutes effective insight-oriented therapy.

And what a rich time for learning it was. The 1970s and 1980s were years of blossoming discoveries for psychotherapists. We were rightfully dazzled by those who came before us: Carl Rogers, who demonstrated the power of a warm and empathetic therapeutic relationship; Fritz Perls and gestalt therapists, who were refining the use of personification as a therapeutic tool; Otto Kernberg and James Masterson, whose theory of ego development delineated the powerful effects of a human's first months of life, along with a careful explanation of how those insights can be applied to help some of the most intransigent psychological problems; Elizabeth Kubler-Ross's research into the grieving process and how it can go awry; James Hillman's descriptions of the interface between Jungian archetypes and psychotherapy, forcing us to step back and reconsider the whole concept of therapy and illness; Irvin Yalom's explanation of existential psychotherapy and how humanity's need to assign meaning to life interacts with a person's psychological well-being . . . and these were just a few of the learning opportunities that embraced me during this period.[1]

None of this rich tapestry of therapeutic learning had much to do with cognitive-behavioral psychotherapy, but Jenny Steinmetz's efforts had been fruitful—I couldn't help but attempt to meld what I had been learning about insight therapy with a strong respect for cognitive/behavioral principles. These attempts met with limited success.

By the early 1990s, clinical research into cognitive therapy had become so compelling that I felt obliged to read more of *their* literature, and I began going to *their* seminars. At first this was difficult. As I suspected, they did not have much of an appreciation for the emotional aspect of human experience and they gave short shrift to the value of insight into a patient's history, minimizing its relevance to his or her treatment. However, there was no denying that cognitive therapists could (and gleefully did) cite a seemingly endless litany of research that demonstrated the effectiveness of their approach. What I found even more impressive than this research was the attitude of some of their adherents: Christine Padesky and Geoffry Young, for example, were perhaps the warmest and most empathetic master therapists that I had encountered since I had attended the seminars of Carl Rogers, the father of "Client-Centered Psychotherapy" in the 1970s. The result of my second foray into cognitive therapy (Jenny Steinmetz representing the first) did not lead to much of an integration of the two therapeutic schools, but I did adopt a variety of cognitive-therapy tools that complemented my insight-therapy approach. And then, satisfied with what I had learned, but still troubled by what seemed like contradictory theories, I continued on, doing therapy, complacently happy about its effectiveness.

This was going to change in the late 1990s as a result of two seminars I attended. I came to these seminars with modest expectations: from one, I hoped to learn a few new techniques for dealing with clients who suffered from post-traumatic stress disorder, and from the other, I expected to pick up a few interesting insights regarding the neurobiology of psychological change. The first was an "Eye Movement and Desensitization and Retraining" (EMDR) seminar. EMDR had been developed by Francine Shapiro, at that time a behavioral psychologist at the Mental Research Institute in Palo Alto, California. Despite its clunky name, EMDR had proven to be very effective in the treatment of post-traumatic stress disorder. (EMDR has since been applied to a much broader range of psychological problems.) I was impressed by how Shapiro condensed a variety of cognitive/behavioral-therapy tools into a few, very effective steps. The second seminar was one that I went to on a whim, partly because I knew the presenter, Jim Grigsby. Grigsby was a neurobiological psychologist doing research at the University of Colorado Health Sciences Center. I had not seen him for a number of years but knew him to be a very entertaining speaker. Also, I was mildly interested in the correlations between neurobiology and psychotherapy. At the seminar my tepid interest evolved into a fascination, and after the seminar I immediately read his book the *Neurodynamics of Personality*, coauthored with David Stevens. I also read the *Neuroscience of Psychotherapy* by Louis Cozolino, and together these books inspired me to discover the rich body of neurobiological research that is cited throughout this book.

I was struck with the possibilities that neurobiology held for informing psychotherapy. However, the research into the area of neurobiology and psychotherapy seemed always to come from a direction that was opposite from my interest. The relevant experiments almost inevitably began by comparing a patient's brain functioning before therapy with how it functioned after successful psychotherapy treatment. By noting the differences, researchers were able to create models that could inform neurobiology about the changes that psychotherapy creates in the brain. For example, a group of people suffering from depression will show an unusual degree of activity in the amygdala regions of their brains. After successful therapy, it can be seen that the prefrontal cortex regions of their brains can now more effectively serve to inhibit this unusual degree of activity. What interested me as a clinician was the possibility of approaching this research from the other direction. Could what we know about neurobiology inform and streamline psychotherapy? The more I learned about brain physiology, the more I began to suspect that neurobiology and a great deal of clinical research dovetailed to a surprising degree, affording new insights into the psychological change process.

I was lucky that at the time I began to delve into the neurobiology of psychotherapy I had also just become familiar with the workings of EMDR, and I was fortunate to have a strong background in insight therapy. These three different perspectives on psychotherapy began to coalesce naturally. When viewed in the light of recent discoveries about brain functioning, many complicated discoveries in psychotherapy research became distilled to their essence and easier to apply. It had long been assumed that that which is psychological is also biological. Now it was becoming apparent that there was a central biological mechanism that underlies much of what occurs in both insight therapy and cognitive-behavioral therapy. Understanding this mechanism points to the possibility of a simple integration of these two seemingly divergent therapeutic approaches. I became convinced that this understanding and its application held great promise for empowering the psychological change process.

I decided that I should put the new concepts to the test: I introduced a new psychotherapy tool to my clients. It was built around a neurobiological model and it borrowed heavily from insight therapy, EMDR, and other proven clinical interventions. I wanted to see how helpful it could be in an actual clinical setting. My previous complacency was quickly shaken as I began to realize that this new way of understanding personal change was very effective for most of my clients, significantly more effective than the therapy that I had previously been doing. Also, as has always been the case, my clients taught me how to further apply what I understood conceptually—how to hone and focus my new formulations. With their help, a new and refined set of tools for psychological mastery coalesced into the Three-Step Practice that is presented here.

It seemed reasonable to me that if a given tool was surprisingly helpful to one rather experienced therapist, perhaps it would be similarly useful to other therapists. And so, at the time of this writing, I have introduced numerous other therapists to the Three-Step Practice, and it is through their encouragement and the encouragement of many clients that this book was written. There are now a large number of people who have contributed to and who deserve credit for the refinement of this approach. It is these people who constitute the "we" that is often referred to in this book.

It is the aspiration of the author and the many people who have supported this undertaking that those who read this book will practice the exercises, thereby creating friendlier, more workable interactions with their own mind and with the minds of those whom they encounter in their daily lives.

ACKNOWLEDGMENTS

Many hands and minds have engendered enormous gratitude for their contributions that made this book possible. Topping the list are my clients, those intrepid pioneers whose courage and insight created the underpinnings for all that this book introduces. Also, I owe a debt of gratitude to the many colleagues who have encouraged and supported me along the way: Marilyn Kruegel, Linda Tuber, Liz George, Dawn Taylor, Gary Groth-Marnat and Sharon Conlin jump to mind. Special thanks to Jim Grigsby, who took time out to read early drafts of the first chapters and then to provide insights that only a neurobiology insider could have. Many thanks to those who have provided editing: Alice Levine and Rivvy Neshema, patient editors of the early chapters who also were invaluable mentors. Special thanks goes to Lizzy Grindey, a patient, competent and generous soul who made the final drafts of this book presentable. Thanks also to Margaret Case for her astute copy editing that made the final product actually look like one. And many, many thanks to Doreen Valentine, who was willing to take a risk on this project and then provide the guidance necessary for the book to become a reality. Thanks also to both of my boys and all of my other family members who for years consistently and patiently listened to my endless reports on progress and setbacks, all the while appearing interested and empathetic—and most of all to Wendy, who read all of the drafts, provided invaluable editing, and shared the sacrifices that this project entailed with playful good humor and support.

What Freud Didn't Know

____ Introduction

Mark was having coffee with friends. As usual, he felt uneasy and anxious, like an outsider who could not effectively join the group. He often guessed that the others were quietly critical of him.

Even though things had been going fairly well in his life, Jerry awoke every morning with familiar feelings of dread and pessimism. It seemed to take almost all the strength he could muster just to drag himself out of bed to face another day.

Jane had managed to alienate yet another friend with her cutting criticism. "Why is it that I, a person who hates being rejected, always do things that result in people rejecting me?" she wondered.

Since Sigmund Freud's time, it has been the job of psychotherapy to address difficulties like those of the people described above. Freud would have attributed their difficulties to internal struggles between id impulses, the ego, and super-ego inhibitions. He would have focused his efforts on buoying up their egos' ability to mediate and moderate in this struggle. Freud's genius becomes evident when we realize that his formulation of ego, super-ego, and id functions are roughly analogous to modern neurobiological understanding of the functions of three regions of the brain: the neocortical regions, the prefrontal cortex regions, and the limbic system. This is especially impressive when we realize how little was known about psychology and the human brain during Freud's time.

In the hundred years that have passed since Freud originated his theories, psychology has dramatically advanced our knowledge of the functioning of the human mind and how to treat its maladies. The process of psychotherapeutic change has, for most people, been reduced from the many years Freud required for successful treatment to a number of months or even a few weeks. Research methodologies that eventually paved the way for this dramatic increase in the effectiveness of psychotherapy took root in the 1940s. During the war effort, the U.S. Armed Services sought scientific answers to a variety of problems, including psychological ones. Psychologists responded and demonstrated that rigorous empirical research could be applied in a variety of ways, including the development of improved treatment tools for psychological problems experienced by soldiers. This resulted in an increased awareness that scientific research methodologies could be productively applied to the study of the human mind.

Because of the interest and money generated by their World War II successes, psychologists in the 1950s decided to apply stringent research methodology to unraveling the question of what constituted the particulars of psychological healing. In the ensuing years, researchers identified many factors that contribute to positive psychotherapeutic outcome. These include the following: working with a warm, caring therapist who portrays accurate empathy; teaching clients the skills of mindful awareness of various emotional states and how to manage those states; using quick and easy methods that promote a client's ability to apply insights about the historical origins of a particular psychological problem; and showing clients how to identify and change the specific patterns of thought that contribute to the continuation of a problem. These and other discoveries generated tools fundamental to today's relatively more efficient psychotherapy.

In this book, we take that research a step further. We are able to present a new therapeutic tool that promises another major advance toward becoming masters of our psyches. This tool takes the form of a simple set of exercises that integrate and simplify a number of psychotherapy interventions that have proven effective. Happily, the exercises are easily learned and usually do not require the assistance of a trained professional. The effectiveness of these simple exercises reflects the fact that our understanding of most psychological problems and our ability to heal them becomes much clearer when we combine the fruits of clinical research with those of neurobiological research.

To understand this, it is helpful to review some recent findings in neurobiology, especially discoveries about the amygdala region of the brain. The amygdala is a small almond-shaped part of the brain that exerts a powerful role in the emotional life of humans. This area of the brain is considered part of the limbic system, and it is the central player in a larger brain network that records memories of painful events.[1] Psychologists sometimes refer to such memories as emotion memories.[2] When an emotion memory arises, the holder of that memory experiences the feelings that he or she was experiencing at the time the memory was recorded. Such memories can have many sources. For example, if a child is teased mercilessly by a sibling, or suffers repeated and painful rejections by a parent, or is humiliated by peers, there is a good chance that those experiences will become the basis for emotion memories. A complex system of brain structures, including the amygdala, mediates this process.

Those regions of the brain that are involved in recording and recalling an emotion memory are often referred to as subcortical, meaning "below the cortex." The cortex, sometimes called the neocortex in humans, is the part of our brain that promotes conscious awareness, paving the way for us to make conscious decisions and to behave in an appropriate, skillful manner. Subcortical parts of the brain are not only anatomically "below the cortex"; figuratively they often operate below the radar of our conscious awareness, and they profoundly influence how we perceive and react to a triggering event. We can say, therefore, that the experiences that arise through the subcortical areas of our brain are activated rather than recalled—that is, they are evoked in a person's mind without purposeful conscious effort and with limited conscious oversight. Emotion memories arise when conditions beckon them, not when we call upon them. All that is required for an emotion memory to arise is a set of circumstances that are somehow unconsciously tied to circumstances that existed when the emotion memory was first learned.

To illustrate this, we could use the emotional feelings that we might experience when we accidentally meet a former lover who has recently rejected us, or when we meet someone who humiliated or betrayed us. Such a meeting can be expected to energize interactions in a variety of areas in our brain. Some of the resulting neurological processing will occur in those regions of our brain that promote awareness and lend support to our ability to behave in ways that are realistic and rational. Other reactions, however, will not be conducive to conscious control. The brain has evolved to include subcortical processing to protect us by preparing us for a flight, fight, or freeze responses. A simple way to understand this would be to imagine two different computers processing our chance encounter with someone who has hurt us. One computer—the neocortex—provides us with conscious awareness and suggests various realistic strategies for handling the situation. The input from the other computer—the subcortical regions of the brain—will be very different. Its influence remains offstage to our awareness. Unbeknownst to our consciousness, this second computer tends to dramatically affect our understanding of the situation and to narrow our behavioral options.

More specifically, in our example, the visual cortex region of the brain mediates an image of a person, allowing other areas of our brain to recognize this person as our former nemesis. This aspect of brain activity engages the consciousness-producing neocortex and seemingly sets the stage for the possibility of rationally choosing to act in a reasonable way toward this person. However, the visual cortex will also mediate interactions between the amygdala and the hippocampal regions of the brain: identifying an object (in this case, a person associated with causing us pain), cueing an emotion memory through the auspices of the amygdala, and activating other parts of the limbic system to release hormones into our body. The emotions that we then experience recapitulate the pain that we felt when we were originally hurt by this person. Our body will automatically ready itself to flee, freeze, fight, or enact a combination of these possibilities. Typically we are only minimally aware of this process, and our neocortex, rather than stepping in to reality-test (for example, "Okay, so we've had bad experiences with this person, but that's history now, and no longer important in my present-day life situations"), instead tends to go along with the reality offered by the subcortical emotion memory (perhaps to the effect, "This person is mean and dangerous so watch out. Be ready to freeze, fight, or flee").

As a result, unless we have learned techniques to master evocative situations such as this, what we experience and how we behave will tend to be at least as influenced by the emotional reaction mediated by the amygdala as by the more rational approach characteristic of input from the neocortical regions of our brain. The subcortical regions of the brain, which include our amygdala, have evolved to promote our survival, and survival takes precedence over more considered reactions. Although these areas of the brain may now seem primitive, they have served to assure our present existence. This is why, although we may feel confused and later regret some of the assumptions and reactions that guided us in "the heat of the moment," our brain is doing the job for which it was designed—directing us to interpret our biased experience of the situation as correct and prioritizing habitual defensive behaviors to activate some combination of aggressive, flight, or freezing reactions.

Brief interactions with those who, like in the example above, have rejected us or otherwise caused us pain in ways common to adult life are usually not especially damaging manifestations of amygdala-mediated emotion memories. Emotion memories such as these, when activated, might cause us passing upset and confusion, but our life will probably proceed relatively unaffected.

In fact, there are many examples of not-so-problematic, scripted emotional reactions that are the result of emotion memories previously recorded with the help of the amygdala. Recently, journalist Jeffrey Goldberg published an article in the *Atlantic Monthly* (July-August 2008) that he called "My Amygdala, My Self." Goldberg depicts his personal adventure in uncovering the workings of amygdala-mediated memories by undergoing a special type of brain scan, while looking at various pictures of people familiar to him, including those in his personal life and those who are public figures. Through the medium of the brain scan, he attempts to deduce his true (unconscious) feelings about the people and issues that capture his daily attention, from his wife, to the 2008 presidential candidates, to television programs like the *Sopranos*. It is all done tongue-in-cheek, but neuroscientists are serious about the potential that these scans hold for understanding how the brain and the mind are related. Brain-scanning tools can be used to delve into peoples' most valid opinions of things (for good or for bad). Goldberg's article and the curiosity it represents is probably a harbinger of a dawning popular awareness about the important role that the amygdala and other subcortical regions of the brain play in our daily life experiences.

Mark, Jerry, and Jane—described at the beginning of this introduction—portray examples of emotion memories so disruptive and problematic to themselves that they chose to seek out a therapist for help. Most of us have at least two or three problematic emotion-memory systems that negatively affect some aspect of our lives in significant ways. What makes some emotion memories cause significant problems for us while others are benign? Usually it is because the most problematic of these memories come from our childhood. This is probably due to the fact that at young ages we are in the process of formulating enduring assumptions about ourselves, about the world, and about who we are in the world. Before our image of who we are becomes more solid and durable, it remains especially vulnerable and malleable. Hurtful events, especially, can then create harmful emotion memories that unconsciously exert undue influence on immediate and later perceptions of ourselves and of the world in which we live. Using Mark, Jerry, and Jane as our examples, we will see that the quality and intensity of their emotional experiences were quite fitting to what they experienced as children, but now, when these emotion memories are evoked, the ensuing feelings, assumptions, and behaviors are inappropriate to the circumstances that trigger them.

Another aspect of emotion memories helps us to understand more fully why such memories can be problematic for us. The hormonal shower that is released into our bloodstream by an emotion memory inhibits the prefrontal cortex, which is the part of the brain that, among other things, promotes our good judgment. Adding this to what we have already noted regarding these memories, we begin to realize the multi-faceted, problematic nature that follows from the activation of an emotion memory: it leaves us with an emotional reaction that neither fits well with our present circumstances nor offers much of a clue as to why it is occurring (it is largely unconscious). It also reduces our tendency to exercise good judgment.

This triple-whammy effect of an active emotion memory—simultaneously affecting our immediate construal of reality and our judgment, yet with so little conscious awareness—leads me to refer to emotion memories as "scripts." When an emotion memory is triggered, it tends to script our experience and behavior. It is as if the subcortical computer we referred to above hands our consciousness (the neocortex) a script that designates how we are to feel and think about an activating situation, and (unless we have learned how to intervene) our neocortex blithely goes along with what the script dictates. We refer to these scripts as "amygdala scripts" because the amygdala plays a central role in a complex process that mediates the storage and retrieval of emotion memories. We know this through extensive neurobiological research on how previously learned, amygdala-mediated emotional responses are unconsciously activated to affect our experience.

When circumstances alert the amygdala and other subcortical regions of the brain to activate a script, our thoughts seem to continue on in somewhat normal manner, telling us what to assume is real and what our options are in light of that reality. These thoughts are, however, strongly influenced by the negative emotions, produced by an amygdala script, that underlie them.

We can see the effect of this underlying emotion component of an amygdala script in the experiences of the people described above—Mark, Jerry, and Jane. For Mark, it manifests in his anxiety; for Jerry, it is his sense of dread; and for Jane, it is the feeling of irritation that she experiences right before she becomes inappropriately critical of someone. Coming into therapy, none of these people understood the nature of why they were reacting in such confusing and hurtful manners.

Seeing that old recorded emotions play an important role in psychological difficulties is not exactly a surprise. After all, a century has passed since Freud first awakened the Western world to appreciate how emotions that had been learned in the past set the stage for psychological disturbances. In doing so, he identified emotions as the energy that drives "neurotic complexes" (Freud's term). The diverse insight-oriented psychotherapy schools that followed in Freud's wake continued to acknowledge the central role that old, previously learned emotions play in most psychological problems.

Freud had no way of appreciating that script-induced emotions (and for that matter all emotions) originate in those parts of the brain that were developed early in evolutionary history. Even now, we are just beginning to appreciate the significance of this. Freud believed that unconscious biographical memories were so painful that the conscious, ego-driven part of the brain worked to keep them safely unconscious.

But a much simpler understanding would prove more accurate. Animals, much lower on the evolutionary scale, have emotional memories akin to ours. For example, a reptile has a limbic system that, when activated by an emotion memory, releases hormones causing physiological reactions that are very similar to those reactions that occur in the human body—reactions that we might, for example, recognize as fear or anger.[3] What is most noteworthy about this is that the part of our brain that mediates human consciousness and higher levels of learning, the neocortex, evolved much later than did the limbic parts of the brain that are associated with the production of emotions. In addition, evolution did not see the advantage in creating a lot of neural pathways between the cruder, amygdala-script-producing parts of our brain and the more refined consciousness-producing parts of the brain. The result is that when an amygdala script is activated, unless we have learned to master it, we have very little conscious insight into what we are experiencing. It is not that our consciousness avoids becoming aware of emotional memories to safeguard us from their energy, as Freud postulated; it is more the case that evolution simply did not spawn the neural pathways that would readily promote this type of awareness.[4]

The antidote to our lack of consciousness regarding what is being activated emotionally for us, whether from a Freudian or more modern-day perspective, is insightful awareness. When viewed from the standpoint of neurobiology, the type of insight most useful in alleviating common psychological problems can be reduced to fairly simple terms: insight that strengthens neural pathways between the consciousness-producing neocortical regions of the brain and the emotion-producing limbic system (especially the amygdala).

This simple understanding of the relationship between the amygdala and the neocortical regions of the brain leads us to look for particular types of healing mechanisms that would serve to induce neocortical areas of the brain to moderate an activated amygdala script. If, for example, we have developed the means to immediately recognize an activated amygdala script and remember that it is a *memory*—probably not a fit for the present situation—then we are more likely to behave and feel in an appropriate manner. In this instance, the neocortex has been activated to rationally evaluate and moderate our reaction. If, in addition, we are able to complement that awareness with an immediate reminder that the emotion memory in question has a particular history associated with it, we have added yet another means by which the neocortex is catalyzed to intervene and moderate our reaction. It turns out that research into psychological change has identified several straightforward and easily learned techniques that readily promote these types of insight. We use these techniques in steps 1 and 2 of the Three-Step Practice described in this book.

How best to apply these techniques brings us to another important difference between what we now understand regarding Freud's concept of "the unconscious" and what we once assumed. Neurobiology has demonstrated that "the unconscious" is more accurately referred to as "unconscious *processes.*" This proves to be an important distinction. Inappropriate emotional reactions (that is, amygdala scripts) are only problematic while they are activated—while the unconscious process is occurring. Amygdala scripts are not a problem when they are not active. More important, attempts to change an amygdala script when it is not activated tend to be fruitless. This is because evolutionarily older parts of our brains (those associated with the amygdala) learn in "real time," while an event is actually occurring. It is most effective, by far, to practice new skills repeatedly while scripted tendencies are running. In this manner, we will learn new responses much more readily, and soon find ourselves functioning in ways that are appropriate and satisfying in those very situations that have historically induced our problems. In psychology such a learning process is referred to as "conditioning" new responses. Old responses can only be conditioned while they are activated.

Freud had no way of knowing that the most powerful way to change most problematic psychological tendencies is to repeatedly condition a new response while problematic patterns of emotions and behaviors (that is, scripts) are activated. Probably due in part to Freud's enduring influence, the world of psychotherapy has been surprisingly slow to embrace this realization. Not understanding this has led to the frustration of many psychotherapy clients (and their therapists) who have gained new insight or who have learned a new way of changing their thought patterns, only to find that the old emotional and behavioral patterns maintain a strong tendency to get activated and run their usual troubling course. Reactions that are initiated by the limbic system are famously unaffected by learning that occurs on a neocortical level of the brain, and so remain separate from and unable to modify an activated amygdala response. To change neurological patterns mediated by the amygdala, it is necessary to activate a response repeatedly and to modify that response while it is occurring. This is the reason that in order to become skillful at those tasks that require sophisticated interactions with our emotions, we have to practice them in environments that evoke the relevant emotions. We learn to handle difficult social situations by practicing social skills with actual people; we learn to be a good teacher by spending time in the classroom with students; we become a good coach by interacting a lot with athletes.

The development of the particular type of insight mentioned above—insight into the emotions that accompany an amygdala-mediated memory—offers an important opportunity to work with old, dysfunctional psychological patterns while they are activated. If we repeatedly induce an amygdala script and then, while it is activated, we practice behavior that requires the neocortical regions of the brain to name and evaluate that emotional response, we can dramatically increase our emotional mastery. This can be accomplished through a technique called "mindfulness" described in chapter 4. Mindfulness has a 2,500-year history in Buddhism. It has recently been adopted and adapted by psychologists in the West and has been shown to be a very effective tool for producing psychological change. The form of mindfulness that we utilize in chapter 4 can be practiced by repeatedly setting in motion a script, and then naming the feeling associated with it. For example, Mark learned that after purposely evoking the familiar feelings of unease and anxiety associated with his script, he could practice focusing his attention on those feelings, noting where he felt them in his body and then saying to himself, "There's that old feeling again." Jerry and Jane practiced a similar technique. What they were learning to do was to identify and objectify the particular emotions associated with their script while the scripted emotions were present. By modifying their internal monologue, pairing a cortical-level response ("There's that old feeling again") with the subcortical emotional response, they were able to begin to moderate the effects of their scripts. Activating and then consciously recognizing the particular emotions associated with a script through naming them is the first step of our Three-Step Practice for mastering an amygdala-mediated script.

Psychology has not been completely immune to the importance of working with amygdala responses while they are activated.[5] Good examples of this are found in a type of psychotherapeutic procedure generically referred to as "exposure therapy." It involves activating a problematic emotional response and, while maintaining that emotional state, conditioning it to have new meaning.[6] Exposure therapy can be readily recognized as a much more refined and focused version of Freudian catharsis. In chapter 5, a particularly effective form of exposure therapy is described and integrated into step 2 of the Three-Step Process.

The stories of Mark, Jerry, and Jane depict another aspect of what contributes to our psychological difficulties. Neurobiology and psychological research have shown that when an especially painful, traumatic event occurs or when a series of painful events occur, an emotion memory is not the only thing that gets recorded. An outline or gist of the circumstances that surround the original events is also created. Later, when an emotion memory is activated, this gist of earlier events imposes itself on present circumstances. Unbeknownst to us (unconsciously), we are compelled to accept a set of assumptions about our present reality that are actually more relevant to our personal history than to our immediate circumstances. Under the influence of an activated script, we experience a felt, but inaccurate, sense of what is. We might feel as if we are a loser, that the world will always work against us, that at any moment we might be found out and crushingly exposed as incompetent. Perhaps we have a tendency to feel as if we are going to be humiliated, or we interpret mildly embarrassing events as truly humiliating. Our friendships might seem unrealistically fragile, or maybe it seems as if our integrity and value as a human being is about to be undermined and so we must defend ourselves. There are many diverse examples of how scripts can make our lives unnecessarily problematic. What seems real to us when a script is activated would appear to an objective observer (especially if he could read our thoughts) as an overreaction or misinterpretation of a situation.

Science has discovered some of the neurobiological processes that contribute to this tendency. Scripts are learned in emotional settings, which in neurobiological terms means that a deluge of hormones is introduced into the bloodstream. These hormones interact with the amygdala to enhance the recording of emotion memories. The hormonal shower is thought to have an opposite effect on a region of the brain, the hippocampus, that creates memory for historical detail. Although there is a historical component to an emotion memory, the historical context is stripped of its detail. This combination creates a memory, unusually vivid in its emotionality but divorced from its historical context. Lacking a context, the memory includes only an outline or gist of what occurred before. Later when that memory or script is engaged, the amygdala and other subcortical regions of the brain again initiate the release of hormones to create the emotion component of an amygdala script. Along with creating an underlying emotional background for our thoughts and assumptions, the hormones also inhibit those areas of the brain, especially the prefrontal cortex, which are critical to reality testing and conscious assessment. The result is that when we are under the influence of a script, present experiences of reality become skewed, taking on the general characteristics of the former reality that occurred when the amygdala script was being created—but with virtually no awareness that those characteristics, now associated with a present situation, more accurately depict a historical situation. While a script is active, it leaves us with the distorted impression that our present situation promises to endanger us with the same pitfalls that befell us, perhaps years ago, when we first learned the script.[7]

We refer to this misrepresentation of reality as the "image" component of a script. Our use of the word "image" is derived from its root meaning, namely, what we "imagine" to be going on. We see this in the experiences of Mark, Jerry, and Jane: Mark's image of himself as an outsider who would never truly be included in social groups; Jerry's image of impending doom; and Jane's image of having to attack others in order to protect herself from being humiliated. (Unfortunately, these images can also become self-fulfilling prophecies.) Each of these images was true to the reality that these people experienced when they were learning their scripts: Mark had been painfully cast as an outsider to his older siblings in childhood; Jerry had struggled in school, getting the message at home that his endeavors would never be good enough; and Jane's comments and attempts to contribute ideas in her family setting were often overridden and contradicted by her disdainful father.

When the activated memory subsides, the distorted imagery that seemed true, maybe only a few minutes ago, might be recognized as an inappropriate or exaggerated reaction. But that realization often comes too late for us to repair the damage that our previous misinterpretation has caused. More often, we do not even notice the distortion that occurs while an amygdala script is activated. Or, worse yet, we defend it by rationalizing our inappropriate behavior.

Freud understood that early painful events, recorded by the brain, may result in "repressed memories" that can, through an unconscious process, distort present realities. "Repressed memories" are a precursor to what we now understand to be the gist or imagery associated with an amygdala script. In the following chapters, we will see that such imagery was a boon to our evolutionary survival—it allowed us to react very quickly to potentially harmful situations, unencumbered by more ponderous conscious evaluations that could have wasted precious seconds when we were facing danger.

Beginning with Freud, psychotherapy has utilized what is often referred to as "historical insight" to moderate the influence of emotion memories. Historical insight provides us with the means to realize that the origins of an amygdala script lie in the events of our past, and are not so relevant to our present. When applied effectively, we recall relevant

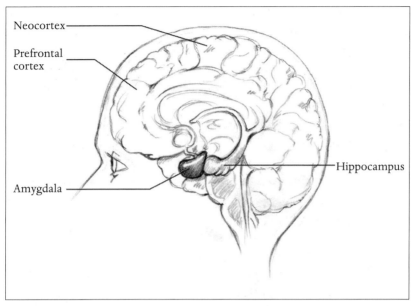

FIGURE 1

Showing areas of the brain central to the recording, retrieval, and moderation of amygdala scripts

Illustration by Kersti Frigell

imagery from the past and pair that imagery with an activated amygdala script, reminding ourselves that what seems true in the moment is, in actuality, an old memory trace. In chapter 5, I describe how psychologists have recently discovered a straightforward way to pinpoint the particular historical situations that represent the seed events that underlie a particular amygdala script. Repeatedly calling upon historical insight while an amygdala script is activated becomes the second step in the Three-Step Practice.[8] Neurobiological studies demonstrate that the prefrontal cortex and the hippocampal regions of the brain can be engaged to moderate an emotion memory when that emotion is repeatedly identified and associated with its historical origins.

After we have discovered the historical origins for an emotion memory, it becomes quite easy to uncover another component of an amygdala script. Imbedded in these images are assumptions about our self or about the world in relation to us. For example, Mark found that imbedded in his image of himself as the perennial outsider there was an assumption that he is not likeable—an assumption that "feels true" when his amygdala script is evoked. Jerry struggles with a belief that he is not able to deal competently with what might arise during the upcoming day, and so, at least while the scripted emotions and imagery are active, he dreads the day. Jane, when her script is activated, sees herself as weak and vulnerable to attack, doubting her ability to protect herself, and so she feels inclined to overreact, vigorously attacking anyone who might represent the type of threat that hurt her in the past. When we know how to recognize the old assumptions associated with an amygdala script and repeatedly practice restructuring those assumptions (while a script is activated), we engage a powerful means to further our psychological mastery. This tool is enhanced if we catalyze that practice with a strong reinforcer, such as the activation of those hormones (for example, dopamine, opiod hormones, and oxytocin) known to be associated with feelings of warmth and well-being. Activating an amygdala script, replacing the assumptions inherent in it, and following this with reinforcement, represents the third step of the Three-Step Practice. This final step in the Three-Step Practice is described in chapter 6.

Each step of the practice is quite powerful in its ability to accomplish the mastery of amygdala scripts. In fact, one can find books written about each step: there are books written about mindfulness-based psychotherapies, books about exposure therapy, books about the therapeutic role that historical insight can play, and books about modifying the beliefs that underlie most psychological problems (see appendix D for an annotated bibliography that includes some of these books). However, by combining neurobiology and clinical insights with a mindful approach, each of the three steps is reduced to simpler and more straightforward forms that allow them to be practiced together. And, in fact, the steps are most effective when practiced together. Although the reader is encouraged to experiment with each step as it is introduced and to get a feel for how each step works, it is not until the reader has been introduced to all three of the steps combined together (in chapter 6) that the steps can be most effectively practiced. Therefore I want to encourage you, the reader, to become familiar with all of the steps and to see how they interact with each other before diligently practicing them.

I and other psychotherapists who have been using an amygdala-script-based approach have found that the exercises presented here are very helpful for a majority of the problems that bring people into psychotherapy, including the most common types of depression and anxiety. We have also found that the concepts are straightforward and the practices simple enough that most people can utilize them without the ongoing guidance of a psychotherapist. However, I want to acknowledge that this book does not purport to address all types and all aspects of psychological difficulties. There are some psychological problems that seem only peripherally, if at all, associated with amygdala scripts—schizophrenia and the more extreme forms of bipolar illness are obvious examples. Also, there are other psychological problems that, while amygdala scripts can easily be seen to play a central role in their creation and in their on-going manifestation, nevertheless require a more complicated and carefully constructed psychotherapy approach. The information presented here includes instructions that allow the reader to decide whether or not these exercises will be helpful to learn by one's self or whether it would be better to seek professional help.

Chapter 8 is written for therapists. In it, I have endeavored to portray how an understanding of amygdala scripts and the Three-Step Practice can be easily integrated into the approaches of many different psychotherapy schools, and I have suggested how to enhance and modify the use of the Three-Step Practice in working with a variety of clients.

An important theme runs throughout this book, one that many readers may find controversial. In this pharmaceutical age, we have come to think of psychological problems as the result of something being wrong with our brain, something that can be fixed with a pill that regulates serotonin or other brain chemicals. There are many parts of our brain that initiate those experiences that we associate with psychological problems, and their interactions are complex. It is scientifically naïve to assume that the simple ingestion of a drug that is not natural to our brain will be able to create the specific changes in this complex system that are required for psychological well-being. Psychoactive medicines can be useful, occasionally even crucial to treatment, but they are often unnecessary, and good clinical practice encourages psychotherapy in addition to any drugs that might be utilized. It is unusual for our brain to have been so affected by an amygdala script that it is structurally damaged. Amygdala scripts are programmed into our brain as the result of earlier painful or traumatic experiences, and they get reactivated unconsciously when we later face situations that evoke those scripts. At least temporarily, this results in potent emotions, imagery, and assumptions about ourselves that obstruct our ability to have fuller and happier lives. Although amygdala scripts are fundamental to much of what ails us psychologically, they are not truly pathological. This is because nothing is broken: the amygdala and its related neurological circuitry is doing exactly what it was designed to do. It was designed to protect us from future hurtful events by preparing us for them. It attempts this by recording and scripting, along interconnected neural pathways, our reactions to situations that have some similarity to the original event.

Freud made wonderful contributions to the Western world's understanding of the human mind, and of course he made mistakes. One of the mistakes he made has had an enduring and unfortunate consequence. By depicting the id and the unconscious as a potentially dark and dangerous area of the human psyche, he misled us into not appreciating and trusting our minds. More recently, many clinicians and researchers have compounded this error through their readiness to label anything that is problematic and painful as pathological or the result of biological anomalies. The result is that we are encouraged to mistrust engaging that which underlies the workings of our own minds. As someone who has spent much of the last thirty years supporting individuals who dare to learn how to mindfully embrace that which arises in their mind, I can say that the human mind is fundamentally an ally to one's self, and that when it is skillfully approached with warm compassion and confidence, we will discover our mind to be a very spacious place that affords us the potential for great freedom—a freedom that allows our humor, insight, and caring to flourish.

CHAPTER 1

Discovering
Amygdala Scripts

To further familiarize ourselves with amygdala scripts, we call upon an expert as our guide: one of those pioneers who has always led the way in the Western world's discoveries about the mind—a client. You will meet Mary and, in a condensed fashion, follow her through her psychotherapy.[1] The unfolding of Mary's therapy portrays the typical stages of people involved in a successful psychological change process. The lens provided by amygdala scripts resolves her experiences into a clear and simple understanding of how psychological mastery can be accomplished.

When Mary came in for her first session, she began by announcing that she had suffered from depression and anxiety throughout her life. However, she had been able to maintain her daily responsibilities. Recently things had gotten worse: "It's been years since I was truly relaxed and I'm not enjoying anything at all. After I called, I realized that I couldn't remember the last time I had a truly genuine laugh and now I'm afraid that I might lose my job." Her boss had told her that her frequent absences at work had caused her to be in danger of losing her job. When not working she would stay at home and avoid contact with people. She also often slept ten to twelve hours a day and even found herself avoiding going out to the commons area of her apartment complex to get her mail. Her doctor had prescribed antidepressants that seemed to improve her mood a little and she felt more energetic, but she still felt very debilitated.

Our first sessions together focused upon identifying those times when she felt more depressed and those times when she felt better. Mary began to realize that there was a pattern to her difficulties.

She noticed that certain situations and thoughts caused a marked increase in her anxiety and depression. She realized that she felt worst when she was around people she didn't know well. Even imagining herself with people aroused anxious and depressed feelings.

————

Most of the psychological difficulties that afflict people can be traced to an aggregate of feelings, thoughts, and behaviors. Sometimes this aggregate is more active and at other times less active or not present. As people begin to track feeling worse and feeling better, one of the first important discoveries they make is that there are patterns to what evokes bad feelings. People who think of themselves as generically depressed, anxious, suffering from low self-esteem, and so on, are often surprised when they notice that difficult behaviors or feelings arise in certain situations and are muted or even absent in other situations. With careful examination, almost everyone can discover that problematic feelings and thoughts are at least exacerbated, and often wholly triggered by something—usually the external events of everyday life or internal thoughts.

A good starting point for working on psychological problems is to take note of what types of circumstance worsen dysfunctional thoughts and feelings. We all have some understanding that our mind continually produces thoughts, feelings, and imagery that guide our everyday experience of reality. Often, however, our conscious attunement to what is passing through our mind is pretty limited, and so we unwittingly allow ourselves to be led wherever that internal dialogue takes us, and it may take us to experiences that are problematic. It is as if we find ourselves on a train that is heading to an upsetting destination and we neither volunteered for the ride nor even, indeed, noticed that we were being taken for a ride. Our best hope is to become aware of the internal processes of our mind so that we can learn to master them. This is why the first step toward accomplishing psychological well-being is to learn how to be more attuned to internal experiences and more aware of the external realities that trigger them. In particular, learning to notice feelings and the type of events that spark them marks the beginnings of practices that can serve to put us more in the driver's seat. The specific exercises that allow us to have increased awareness and control over our internal narrative are called "mindfulness practices." In chapter 4, we describe a mindfulness practice that allows one to focus the attention on the nature of what brings on their uncomfortable feelings. By recalling those situations that have led to problems in the past, a person can evoke them in therapy sessions or self-help practice sessions and then work directly with the evoked feelings and thoughts. This creates a less abstracted and more true-to-life environment, where the feelings can be worked on experientially and effectively.

––––––––

Once Mary began to notice more about the situations that activated her discomfort, she also realized that she had been attempting to avoid them for a long time. She had developed various rationalizations for her avoidance behaviors, claiming, for example, that "People are so involved in their own view of the world that they won't understand me and will be critical of me unless I pretend to be someone I'm not." As she focused upon the times when she felt worst, Mary came to see that compelling feelings and thoughts were the underpinnings for her discomfort. She began to realize that she had been only vaguely conscious of her real motivations for isolating herself. She discovered that her efforts to avoid people were really attempts to avoid painful feelings.

Mary also began to identify many habitual thoughts and behaviors that she had adopted to distract herself from feeling bad. For example, she would often preoccupy herself with romantic daydreams, and she almost always had the television on.

Not surprisingly, Mary was quite taken aback to realize how much energy she had been putting into avoiding painful feelings and thoughts. But she was also a bit pleased by her discovery because she could now see much more clearly and specifically what she had to work on—namely, bad feelings that were being evoked by certain kinds of thoughts and situations.

––––––––

Painful feelings are difficult for all of us but they are especially confusing and upsetting to children. We have seen that amygdala scripts include uncomfortable emotional feelings, and it is in childhood that we learn most of the scripts that become problematic to us as adults. In an ideal world, all children would learn to relate to painful emotions by talking to an adult who is a willing and skilled listener. Such supportive conversations would encourage scripts to remain more available to consciousness, making them more amenable to modification. Unfortunately, adults adept at empathy are not always available to children, and even when they are, children often opt to avoid uncomfortable feelings rather than talk about them.

Behaviors that allow us to distance ourselves temporarily from disturbing feelings are compelling—momentarily dulling the intensity of the emotion component of a script—but there is a cost. What worked once to distance us from the impact of a disturbing emotion tends to get repeated, and repetition leads to the development of habitual avoidance strategies that operate under the surface of our awareness, making it more difficult to identify and work with a script. Without being aware of what we are doing, we keep using the same well-practiced defenses to squelch the pain of a script. Because the emotions, imagery, and assumptions that are inherent in an activated script do not fit one's immediate circumstances, our efforts to avoid them seem similarly incongruent with our situation. While the underlying script may not catch our attention, our efforts to protect ourselves from a script often do. In fact, it is often the dysfunctional, defensive maneuvering that drives us to engage and work with a script, not our awareness of the script itself.

The habitual defensive mechanisms that we utilize employ versions of the fight, flight, or freeze mechanisms. For example, we act as if it is less painful to be quick to anger (fight) than it is to re-experience a past rejection that has become part of a script; less painful to withdraw into ourselves than to hazard being exposed and hurt (flight); less painful to not talk to anyone than to risk the possibility of criticism or rejection (freeze); and sometimes it may even seem less painful to be buried in a fog of depression or preoccupied with anxious thoughts than to hazard further criticism and rejection (a refinement of the "freeze" response). The type of avoidance strategy that a person adopts seems to be partly the result of circumstances and partly the result of genetically determined personality traits.[2] Mary's increasing tendency to isolate herself, combined with the lethargy of her depression, served to keep her removed from the pain associated with her underlying script. But it was not the script that caught Mary's attention; it was her defensive behaviors that brought her into therapy.

Habitual behaviors that serve to protect us from more uncomfortable underlying thoughts and feelings can take fairly extreme forms. In Mary's case, these behaviors were: isolating herself from almost all contact with people outside of work, sleeping excessively, and preoccupying herself with continual daydreams. More often, hiding from scripts takes more benign forms, such as habitually avoiding only certain people that make us uncomfortable, or not being in crowds, or circumventing conflicts in unproductive ways, to name a few.

Some defenses are passive, such as trying to avoid evocative situations, while other defenses, such as being quick to anger, would represent more active maneuvering. Other examples of active defenses might include claiming the center of attention, being too agreeable, habitually showing off our intelligence, or feeling like we always have to win an argument—the examples are many and they are all costly. These "symptoms" are usually what first alert us to the presence of a script.

Once we understand the nature of our defensive responses, they can become helpful to us. Noticing the habitual ways in which we defend against a script can represent a wake-up call to recognize that script. Our mechanisms of distraction, while costly, are nevertheless fairly effective—so effective, in fact, that it becomes easier for us to notice the seemingly odd things that we do to protect ourselves from old painful feelings than it is for us to notice the painful feelings themselves.

Freud was the first in Western psychology to describe habitual defensive maneuverings as a strategy to safeguard against unconscious, painful emotions. His understanding of defenses remains largely intact to this day. However, modern psychotherapy has come a long way since Freud. In the decades that followed Freud, psychotherapists have developed straightforward methods that help people to readily identify and effectively address their defenses.

The mindfulness practice that we mentioned above is a recent and powerful example of those methods. Taking note of those situations that seem difficult to us, or annoying to others, promotes our ability to recognize and engage our scripts. Importantly, mindfulness is only helpful when it includes a compassionate attitude toward oneself. The discovery of an undesirable habitual defense should be approached with openness and nonjudgment. Everyone wants to avoid bad feelings. When we unconsciously act in unfortunate ways to avoid distressing emotions, no matter how costly or unattractive our behavior may be, we can still adopt an empathetic attitude toward ourselves, recognizing and appreciating that our unfortunate behaviors are simply an attempt to avoid pain. What could be more natural and understandable than that? Our willingness to look at this and acknowledge it is an act of courage. Cultivating fearless compassion for ourselves lays the groundwork for daring to discover and engage amygdala scripts.

Lessons derived from mindfulness practices include a psychological rule of thumb: if we want to study something that has been psychologically difficult for us, we should start by paying attention to concrete examples of the times when the problem occurs. Mary applies this by noticing those specific situations that sparked her problematic behavior. It was a careful examination of actual situations that led Mary to discover her particular repertoire of scripts.

Any of us may do the same. We may, for example, begin to notice that we have a tendency to attack and criticize others. When we look carefully, we can see that such habitual attacks are always enacted toward someone who is doing or saying something that threatens to (or actually does) activate a painful script. For example, we might overreact to people who disagree with us or we might overreact to people who are unusually forthright. If we notice this tendency, then we can consciously recall a situation where it occurred. Upon careful examination we might realize that we have a habitual, learned fear that we have previously been only vaguely aware of—perhaps an unrealistic or exaggerated concern about appearing stupid in front of others. When this reaction is on the verge of surfacing—for example because we are with someone who is not shy about disagreeing with us—we are given an opportunity to consciously discover costly, unconscious tendencies. Or, to cite another example, maybe you notice that you respond to passive people with undue irritation and criticism. In purposely recalling specific instances of this, you might be able to uncover distressing feelings and a hidden agenda to avoid them; perhaps by being critical of people who act submissively, you are reassuring yourself that you will never again have to experience the humiliation of the passive victim you once were.

Another example of defensive behavior is an aloof response—being unaffected by things. It is a strategy of being "thick-skinned," but it tends to result in being "out of it." It leads to seemingly being unaffected by events that would normally be expected to have an impact, such as a spouse's angry rejection, the loss of a friendship, or even the death of someone close to us.

Miring ourselves in a loop of self-negation is, oddly enough, another form of self-protective behavior that was originally learned to avoid the pain of an amygdala script. A preoccupation with our inability to transcend our perceived failings creates a mental loop of hopelessness and helplessness that keeps us one step removed from the more painful underlying hurt we once suffered. It is somewhat less painful, and it is easier, to get caught up in mind-numbing habits of self-recrimination than to feel the raw awfulness of childhood abandonment or rejection. Habitually revisiting self-criticism may also, at first, be experienced by children as vaguely reassuring, suggesting the possibility that maybe, if they remind themselves how limited they are, this will someday help to get it right, so that they won't have to feel the misery of their pain. For example, children who are learning to deal with a critical or rejecting parent or sibling will often lapse into a pattern of self-negation, thereby avoiding the sting of the immediate rejection while also keeping alive a forlorn hope that if they ever become the better person that they "should" be, the painful rejections will cease. These habits of self-criticism, it turns out, are the cognitive habits that have been shown to underlie most depression.[3]

Becoming aware of these unconscious habits edges us closer to what lies beneath—closer to being able to work with the components of emotion, image, and belief. At this point you might find yourself saying: "Whoa, just a minute here! There might well be some pretty good reasons that I don't want to re-experience those painful things that I've been avoiding all of these years." And unless you have effective tools to master these things, you may be right. However, the very good news is that all three components (the emotion, image, and belief components) of a script can be worked with effectively. This is not just wishful thinking. For over a hundred years, therapists and psychologists have carefully researched how to deal with each of the three underlying components. As a result, these researchers and therapists have honed more and more effective ways of dealing with them. Recently, psychotherapy research has been joined by discoveries about how the brain functions to augment our knowledge of the psychological change process dramatically.

But we are getting ahead of ourselves. Let us return to Mary. Her story can help us further understand scripts, how they relate to brain functioning, and ultimately how we can effectively work with them.

———

Mary continued to examine the situations in which her script was activated. I suggested to her that as she imagined herself in those

situations, she focus upon her physical sensations. Mary began to
notice that there was a sick, butterfly-like feeling in her stomach and
a sad, broken-hearted feeling in her chest. These feelings seemed to
be at the core of what she had been trying to avoid. Often these
feelings were quite intense. Even so, by repeating the process, Mary
found she could tolerate invoking those feelings. She also realized
that these feelings had haunted her most of her life. It was a relief to
discover that she no longer had to fear them.

———

We tend to think that emotions originate and exist separately from
our bodies when, in fact, various physical sensations are the experiential
basis for emotions. The particular sensations that arise in our bodies
when a script is activated represent the emotion component of a script. If
a script has an emotion component that includes anger, we might notice
a characteristic feeling of energy and tenseness; if the emotion component is anxiety, we might notice that we feel a particular "ugh," or but-
terflies, or other discomfort in our stomach area. Other scripts might
engender a sense of sadness or a pang in our chest area. Especially at first,
people might only notice that various parts of their body tense up when
something is activated. We will see in chapter 4 that, with a little practice,
it becomes easy to distinguish between those emotions that are caused by
a script and those emotions that are appropriate to our situation.

———

Mary's ability to watch the emotion component of a script and not
get entangled in it was enhanced by evoking the emotions and then
stating, "There's that feeling again." She told me: "Feelings that I've
been avoiding all of my life, now seem familiar and easy to tolerate.
I'm also realizing that the feelings that arise when a script is active
don't really fit the situations that trigger them."

———

Mary furthers her mastery of the emotion component of a script
through practice sessions. She begins these by imagining those situations
that provoke the emotion, imagery, and assumptions associated with the
script she is working on. She then focuses her attention on the feeling
component of that script and, once recognized, she tells herself, "There's

that feeling." This is good news, for Mary demonstrates a fact that we know from direct observation of human brains: brains are very trainable, or as scientists like to say, brains are "plastic." When we practice even simple exercises, neurological pathways in the brain become altered: areas of the brain that would formerly not show much activity now become activated. We purposely activate old scripted feelings by calling up recent events that evoke them. Then as we consciously identify and name the physical sensations associated with the emotion component of a script ("There's that feeling"), the prefrontal cortex is engaged and new awareness arises through the activation of other areas of the neocortex. The conscious part of Mary's brain, becomes increasingly less overrun by the amygdala-mediated memory (a script).[4] By repeatedly evoking a script, noting the sensations in her body that are associated with the emotion component of that script, and then verbally acknowledging that component, Mary is reinforcing the activation of neural pathways that allow the prefrontal cortex areas of her brain to moderate an amygdala-mediated response.

———

After practicing noticing and naming the emotions associated with her script, Mary ran into an obstacle. She was still bothered by her perception that people were critical of her. This seemed especially true when the script was activated. She reported, "When I'm out in public or having lunch with my coworkers I feel very awkward and I think that people are put off by me. My mind then begins to remember all of those times when people have been critical of me or seemed reluctant to be with me, and I begin to focus almost all of my attention on things in the present that seem to suggest that I am an awkward dolt. It becomes almost impossible at these times to not get overcome by the awful emotional reactions I've been working on."

I suggested that we take a closer look at what seemed to be happening on these occasions. As she examined this over the next couple of sessions, Mary began to suspect that these old familiar images of herself in relation to other people might not be as accurate as she had previously assumed. "When I notice that old scripted feelings are active and present, I try to evaluate more carefully what is going on. I'm seeing that people are not nearly as critical of me as they once seemed. I'm also realizing that when I do put people off, it often starts when I think that they are going to not like me so I get defensive and I talk too much. I act like I expect them to be critical of me. I've also noticed that sometimes people are being critical of me, but I've begun to notice that I'm not the only one—many other people get criticized too, and I don't have to be so sensitive about it. So what if a few people don't like me? It's not going to kill me."

Mary discovered the second component of the script—the *image* part. As we mentioned in the introduction, what we mean by image here is not primarily a visual image, but is instead what we *imagine* to be true. When her script is activated, Mary experiences an image of what is happening that does not accurately reflect the reality of what is happening. Whereas the emotion component of her script suggests to Mary that she is in danger of being hurt, her image component depicts in a general way the "how" of the hurt—people she knows are going to secretly dislike and discount her. The image component of a script transposes onto present circumstances an inaccurate general outline of what is real. A script imposes assumptions into a person's internal narrative, so that the neocortex becomes subsumed by a script and interprets present reality in terms of the gist of the painful circumstances that were present when a script was originally being learned. Usually, when describing this part of an activated script, people start sentences with "It seems like . . ." (for example, "It seems like people only care about themselves and couldn't care less about me or others"). Such exaggerated or inaccurate images of what is real take a variety of forms. It might seem as though situations are more dangerous than they are; or it might seem to you as though no one will ever truly care about you; or seem as though no one will want to hire you; or seem as though if you make a mistake everyone will realize that you are incompetent, to name but a few of the almost innumerable forms that an image component might take. Another common image includes feeling unrealistically responsible for the emotions and well-being of others, leading to the overwhelming and burdensome sense that you are in danger of letting someone down. Examples of scripted imagery are many, varied, and common.

Sometimes people know, in retrospect, that their image of what was going on didn't entirely make sense. However, they still might find it difficult to shake "what seemed true." The next time they find themselves in similar circumstances, they are again overrun by the same image. Even when people have the insight that a script is presently activated, they may still find it very difficult to reassure themselves that what seems true may not be. In the following paragraph, we will see how Mary learned a very effective technique that helped her further tame the images of her script.

———————

At home, Mary had been repeatedly activating the feelings associated with a particular script and then practicing the first step by saying to herself, "There's that feeling." Now I instructed her to evoke the feelings again, but this time to remain actively focused on those feelings in her body without making any statement about them. She soon noticed that as she remained focused on the feelings in her body, her mind would naturally begin to drift from image to image of times in the past when she had had the same feelings. It was like a slide show. Recent images of rejection came to mind, followed by older and older images: images of being laughed at in middle school, of not being invited on dates in high school, and finally images of her father's criticism and her alcoholic mother's emotional detachment from her. I asked Mary to choose one of the earliest images, one that would depict the kinds of painful circumstances that had left her with the emotion memory. Without hesitation, Mary closed her eyes and described a memory: "My Mom has just left the room after fighting with Dad. They had been yelling at each other but I don't remember what they were saying. After my Mom left, my father sat me down and said that my Mom was the way she was because I was such a worry to her. He said that I had to be very, very good so that my mother wouldn't be like this. I was about five years old at the time." We decided to use this as a seed image—an early image that epitomized how Mary had learned the script she was working on.

We discussed how Mary could use this image to add another step to working with her script. After thinking of recent difficult situations that would evoke her script, noticing the feelings in her body and saying, "There's that feeling," she would add a second step by reminding herself of that seed image and then say to herself, "That's where I learned these feelings." An important confidence-builder was Mary's newfound understanding that the feelings being activated made sense; these were simply feelings that she had learned long ago. She began to realize that her mind was not broken or damaged; it was unconsciously calling up old memories. The amygdala part of her brain was doing one of the things that it was designed to do: mediate the storage and recall of old painful memories.

———

There are various effective ways to work with the image component of a script. These methods usually involve gaining a clear idea of the historical situation where the problem began. Through practice—evoking the feeling and image again and again, and applying corrective insight each time—you can become very effective in allaying the problematic effects of the image part of the script. This corrective insight is the statement, "That's where I learned it." (I suggest that the reader not yet practice this step without the more detailed understanding that is described in chapters 5 and 7.)

———

During the session after identifying a seed image, Mary complained that what had previously been smooth progress toward the mastery of her script now seemed to be bogged down. "When I evoke the feelings of this script and then remember the situation of the seed image, the emotions are so strong that I find it very hard to stay with the practice."

———

Since all scripts have an emotion component that represents the energy that has been driving the script since its inception, it is natural to experience strong emotions while working on a script. It is also not unusual for those emotions to be intense enough that it makes it difficult to effectively practice with that script. This is especially likely to be the case when a person is working with seed imagery. Fortunately, psychology has developed a very effective means by which emotional intensity can be reduced.

———

I suggested to Mary that she use some recent event to evoke the feelings associated with her script and then imagine the seed situation that she had discovered the week before. While maintaining her awareness of the emotions associated with that early situation in her life, I asked her to keep reminding herself that those strong feelings were learned back then, that those feelings no longer have relevance to her present life, and that they are simply an emotion memory of something that happened a long time ago, and so on. During the session Mary continued to do this exercise for a little longer that half an hour, tuning into the seed image and how she felt then, while reminding herself that she is now in a different place and time, and that the emotion memory is an artifact from the past. As she did this, Mary was very pleased to notice that these old familiar feelings were losing much of their intensity. They became less intense during the session and also subsequently during her daily life.

––––––––––

The kind of work that Mary was doing by imagining the original "seed situations" and maintaining an awareness of her emotions while reminding herself that those emotions were simply an activated memory is a form of "exposure therapy." Another similar technique has a more "touch-and-go" quality. People call up the feeling, acknowledge it as an old feeling, and then let it go by returning their awareness to their present situation. (These practices and others are detailed in chapter 5. Please do not practice this step without the more detailed explanation that is described in chapter 5.) Exposure and "touch-and-go" exercises allow us to practice engaging emotions and to use our awareness to reduce their intensity. These are two of the most common means that therapists have found to effectively reduce the intensity of problematic feelings.

––––––––––

Mary's behavior and life had begun to change in important positive ways. For example, she noticed that it was becoming easier to be out and around people. Also, by repeatedly applying her new historical insight—connecting the present-day scripts to their origins in the seed imagery from her past—Mary was able to remind herself that what was truly happening when her script was activated was that old feelings and images imposed themselves onto the present reality and that the old images probably did not fit in her present life. She became reassured that being out among people was not so fraught with the danger and humiliation that she had always secretly feared.

———

So what was left for Mary to learn?

———

While Mary was focusing on an early image during which her parents made her feel bad about herself, I asked her, "At the time of the seed image, when you were a little girl, what was the message that you were being given about who you were?" Mary quickly realized that at that time she had learned to think of herself as a bad person, unworthy of love, and prone to make those around her unhappy and irritated. Seen through the eyes of the child who was in the seed situations, these assumptions were reasonable.

———

When an emotion memory is activated in us, if we look carefully, we can see that along with the emotions and the image, there is another important component of what is going on, one that is implied in the image part but that needs to be isolated and worked with. This part is the belief component. With a little bit of practice, you can notice that inherent in an image that is activated are beliefs about yourself, or about the world. If you are depressed, and you observe yourself very carefully, you will probably notice that you have a tendency to feel bad about yourself, and to think of yourself in self-critical terms: "I'm pretty messed up," "Of course, no one will like me—I don't even like myself," and other similar self-negating beliefs. Or if you are anxious, you might be able to see that when the anxiety is activated, you have frequent thoughts that something bad is about to happen—and you are helpless to protect yourself. In fact you might think that you are a rather helpless person, someone who is prone to falling apart. Maybe you fear that you are in danger of going crazy, or that you are about to die, or that you are the kind of person who is prone to self-humiliation. All of these and many other similar beliefs are powerful inducers toward getting stuck in a bad feeling.

———

Mary realized that it would be much more helpful for her to let go of the negative beliefs and instead to believe the opposite about herself: that she was a good person, worthy of love, and able to be helpful, even a delight to others. She and I worked with this new belief in various ways until it "felt" true. Mary then began applying the new belief to herself as she imaged herself in the old seed situations, and as she imaged herself in various situations in her present life. It was especially helpful to practice applying the new message about herself as she imagined those situations that historically had felt so awful.

Mary found that her life had changed a great deal. It was not that the old script never got activated. It did. What was different was that when it was activated, Mary immediately recognized it as a script and reminded herself of her new beliefs about herself. The components of the script were of historical relevance, having no valid place in her present reality.

Here's some more good news. Psychologists have developed especially powerful tools to help with the belief aspect of psychological problems—the belief component of an amygdala script. This approach to psychotherapy is referred to generically as "cognitive therapy." Because they are so effective, cognitive approaches have dominated the most rigorous psychotherapy research journals since about 1990. Cognitive therapy involves learning to identify the particular thought patterns that create or exacerbate a particular problem, to stop those thoughts, and then to carefully replace them with other thoughts that are affirming. This step of the Three-Step Method is referred to as the cognitive change step. (Chapter 7 provides detailed information about how to best utilize what has been developed through cognitive therapy research.)

At this point an odd thing began to happen to Mary. Becoming more masterful with these scripts seemed to evoke in her quite strong feelings of sadness. I encouraged her to use mindfulness techniques to tune into the origins of the sadness. She noted that these particular sad feelings did not seem like old scripts. They felt more like regret about her life. She realized that she was grieving. For many years her scripts and her defenses against the scripts had limited her, and she had missed out on a lot. She was grieving for what she had lost as a result of her script. In therapy she would sometimes cry, saying, "If only I had known this before."

Mary continued to practice working with her scripts until she became masterful with them. When an old script was activated, Mary could now quickly recognize it and go through the three steps: identify the feelings in her body that were associated with that script; state her insight into the image component of the script by reminding herself where she had learned the image; and then change dysfunctional beliefs about herself by replacing old beliefs with more accurate affirming beliefs.

Occasionally, in an especially evocative situation, Mary had to repeat the three steps, but usually the whole practice (from step 1 through step 3) only took about fifteen seconds. And the result was that the script's control over her dissolved, leaving her unencumbered and free to relate skillfully to her present situation.

———————

Through the Three-Step Practice, clients like Mary leave therapy with practices that will continue to benefit them. The three steps bear repeating: (1) identifying the feelings associated with an activated script (mindfulness), (2) making sense of the feelings through remembering where the script was originally learned (historical insight), and (3) changing the erroneous beliefs that were learned at that time (cognitive change).

These steps are presented in much more detail in the following chapters. Once the steps are learned, practicing them for even only twenty minutes a day can very effectively help you to become mindful of and masterful in dealing with a psychological script that might otherwise haunt you for a lifetime. When a particular script has been mastered, all that is required is to remain on the lookout for that script, to identify it when it is evoked, and to apply the three steps described above. The practices may seem complicated at first, but hundreds of people—many of whom are presumably not unlike yourself—have learned and benefited from them. The next chapters are carefully designed to ease the way to understand and master your scripts.

CH

Amygdala Scripts

A CLOSE LOOK

The concept of amygdala scripts is drawn directly from research in neuroscience. When we look at psychotherapy through the lens of brain science, what we find is not merely intriguing—a neat convergence that shows what we have learned about psychotherapy has identifiable correlates in the brain—but more importantly, it is surprisingly suggestive about how we might improve psychotherapy. In particular, neuroscience points to ways we can concentrate and target the psychological tools that have become the mainstays of psychotherapy. Furthermore, the enhancements to the psychological change process that are provided by an understanding of amygdala scripts are not complicated. Individuals can readily learn to apply the resulting theory and practices and can often do so without having to engage in psychotherapy.

Amygdala script theory is made possible by recent advances in scientific research. For example, in the past ten years, Magnetic Resonance Imaging (MRI) studies and other imaging techniques have allowed us to observe neural function under various conditions with detail that could only be dreamed of a few decades ago. At the same time, researchers have continued to study the functions of the brain in the more traditional way, using improved methods for pinpointing the location of lesions and measuring the dysfunction that results. These newly refined research tools have provided an increasingly detailed understanding of how the brain functions, yielding knowledge that would have astounded the scientists of Freud's time.

Neuroscience does not stand alone in its new insights into how the mind works. The world of psychotherapy is also abuzz with discoveries. During Freud's time, a few clinical examples could suffice as the evidence necessary to launch a whole new theory of mind. Now improvements are grounded in carefully designed studies. As a result, in the past few decades researchers have made important progress in addressing a series of questions about psychotherapy, beginning with "Does it really work?" (yes)[1] to "What types of psychotherapy work best?" (debatable)[2] to "Which components of a particular psychotherapeutic approach underlie its effectiveness?" You might accurately guess that the answers to the last question are fairly complicated, but we can now point to a number of components that have been shown to be especially effective in promoting psychological change: mindfulness,[3] insight,[4] cognitive change,[5] and behavioral techniques,[6] to mention a few.[7] It is as if we can finally open the hood of the psychological change vehicle and examine the mechanisms that make it work.

Unfortunately, we have lacked the interdisciplinary awareness necessary to realize the great potential for cross-fertilization that these breakthroughs promise. Insulated from what our colleagues are doing, clinicians and the scientists who support them have remained largely unaware of the connections between their work and the work of researchers in brain science.

There is another reason that we have not combined existing biological and psychological knowledge, and it can be traced to Sigmund Freud himself. The doctors of the early twentieth century, including Freud, were charged with curing disease. Their approach assumed that diseases are caused by a breakdown in our physiology—something is broken (pathological), it is not functioning the way it was designed to function, and so it needs to be fixed. That the origin of psychological problems lies in a diseased mind is a concept that has continued into the present and has served to cloud our understanding of biological research. The prevailing logic seems to run something like this: if psychological problems are painful to people, appear in diagnostic manuals as illnesses, are covered by insurance companies, and can sometimes be helped with drugs, then they must be caused by a damaged or abnormal physiology. In searching for physiological correlates to psychological problems we have been looking for brain anomalies. Although we have found patterns of brain functioning that seem to correlate with psychological problems (which are cited throughout this book), we have found scant evidence of major structural problems. Nevertheless, the bulk of neuroscientific study continues to seek physiological anomalies in the brain, overlooking what is being more and more strongly suggested by that research: healthy brains can create dysfunctional psychological patterns.

Amygdala-script theory appreciates the strong evidence which suggests that the brains of most of us who show up in a psychotherapist's office are not damaged or abnormal; in fact, our brains are doing exactly what they were designed to do. The problem lies in the fact that some of the tasks for evolutionary survival that the human brain has been designed to do, and now does so well, are no longer helpful. The natural tendencies of the brain serve to create painful and unhelpful psychological responses. They are like tonsils that in doing their job tend to get inflamed and cause problems. Unfortunately, unlike tonsils, problematic parts of the brain (the amygdala being a prime example) cannot be removed without impairing other important and very useful functions that these areas of the brain carry out on a daily basis.

Any change that occurs in our minds (for example, learning about amygdala scripts from this book) has a neurological correlate—something changes in our brain. So if individuals are suffering from depression or anxiety caused by an amygdala script, then we can expect to see that their brains function differently from those who are not experiencing those difficulties—especially while their symptoms are most pronounced (when their scripts are activated). Too often the media, and sometimes even researchers, seem willing to accept the incorrect assumption that just because a psychological difficulty has neurological manifestations it is therefore not amenable to change. This is simply not scientifically valid. In the last decade or so, we have learned that the human brain is surprisingly plastic, amenable to change. In some of the more extreme manifestations of psychological problems, there are neurological processes that, even if their origins are a result of repeated learning, they are sufficiently mutable that a combination of long-term psychotherapy and pharmaceutical interventions can treat them. But these more extreme cases that require long-term treatment are the exceptions. Volumes of psychotherapy research underscore the reality that the brain can readily change in ways that promote psychological well-being. Although there may be identifiable differences that make a few of us especially vulnerable to the traumas and pain that generate amygdala scripts, most of these idiosyncrasies are not inherently a problem. We may have a sensitive nature, for example, making us especially sensitive to our own feelings and to the feelings of those around us. Especially as children, this could make us vulnerable to more strongly felt painful experiences and so more vulnerable to the creation of scripts. This same characteristic could also serve, however, to make us especially insightful, empathetic, and helpful to others.

When we stop seeing psychological problems as necessarily resulting from an abnormal brain, we discover how a perfectly healthy brain can learn to behave in very problematic ways—that is, a well-functioning brain can lead to dysfunctional behavior. Most psychological problems, while painful and harmful, are accurately understood to be dysfunctional rather than inherently pathological. If we cease to look at our problems as the result of something irrevocably broken in our brain, we are free to consider a new perspective, one that provides a clearer understanding of what ails us psychologically. This new perspective has a markedly salutary effect upon those of us who are trying to change. It takes courage and discipline to undertake a process of personal growth. It is not helpful to add an unnecessary burden to that process by suggesting that our brains are damaged.

To better understand this perspective we can look to the evolutionary advantages offered by the amygdala and other related areas of the brain. Imagine us as our early ancestors, on a typical day, foraging for food in the savannah. Our attention is drawn to an unusual pattern of movement in the tall grass surrounding us, and then a flash of orange, and we are horrified to see a tiger brutally kill someone in our group. Our experience of this traumatic occurrence involves a complicated reaction in our brain that includes the amygdala and the recording of the event into what we are terming an amygdala-mediated emotion memory or amygdala script. It is the nature of this memory that, if ten days, ten weeks, ten years, even thirty years later, we see a similar cue—movement in the tall grass perhaps, or a suggestion of orange, or some other detail that is associated with the original learning situation—our bodies will react immediately. Our pulse rate will go up, our skin will moisten, our muscles will tense for action, and our digestive processes will be put on hold. We will become automatically hyperalert and hyperattuned to any other conceivable clues that might indicate danger. The previously traumatic incident has been stored in our primate brain as a particular type of memory, and this memory has now been activated. Once the memory is activated, our tendency is to *feel and think as if we are in danger* of a repetition of the original painful situation. We even begin to look for reasons to support this assumption—something awful is about to happen, so we had better be on the lookout to protect ourselves from whatever it is.

Imagine for a moment that our evolutionary ancestors had to deal with survival without these types of memories. This would mean that human memories of hurtful and traumatic experiences would take the rational, nonemotive, more analytic form that characterizes the neocortical parts of the brain (the part of the brain that is in charge, for example, when we are working on a crossword puzzle). If instead of amygdala-mediated memories, the tiger attack had generated only the higher-level kind of memory that we usually think of when we think of memories, then, a few years later when we are out on the savannah and we see the rustling in the grass and another flash of orange, we would, in a very conscious evaluative way, assess things. Our only reaction might be to think: "Well, that flash of orange might be another tiger, but there hasn't been a tiger attack for years. The orange could also be those poppies that I noticed growing over there the other day. And I noticed that Jean was wearing an orange top this morning and the angle of the sun is such that it could be the sun glinting off of her top." We can easily imagine the effect upon the tiger population that such a cerebral and more nuanced, strategic approach might engender. Word would spread in the tiger world: "Hey, there is a great source of protein out there on the savannah. They're a bunch of ape-like animals and when you approach them, even if they glimpse you, they stop and ponder for a moment or two, giving you a chance to pick and choose which one you want to eat!" In order to better survive, when confronted with the possibility of harm, our species continued to employ an amygdala-mediated response. The more rational, analytic part of our brain has been most useful as a survival mechanism primarily when we are not facing what might be an imminent threat. Although more realistic, analytic reasoning might occur, it is vulnerable to being overshadowed by an activated amygdala script. We may be attempting to think clearly, yet our body is signaling danger. Our brain is structured to prioritize survival, and the evolution of our brain attempts to accomplish this by allowing higher reasoning to be inhibited or even overrun by subconscious concerns that also manifest in our bodies as uncomfortable emotions

Let us take a closer look at this situation from the perspective of memory.[8] When we think of memories, we usually think of the kinds of memories that researchers sometimes refer to as "explicit" memories, involving the conscious recall of historical events. Many kinds of memories exist, however: kinetic memories of practiced bodily responses (for example, we remember how to ride a bike or how to keep our balance while skiing down a hill); sensory memories (we can recognize a fruit by its smell); visual memories (we recognize a friend); and many others.

The type of memories that concern us are sometimes referred to as emotion memories. They are unconscious memories of painful events and they are the basis for amygdala scripts. This type of memory has its own peculiar characteristics that exist to protect us from hurtful, threatening experiences. The function of an amygdala script is to respond to those cues which suggest that something hurtful might be immediately pending and to then activate avoidance strategies. Amygdala scripts represent neural patterns that are not relevant to the more nuanced niceties of our existence; they are only concerned with survival. Once activated, amygdala scripts are not effectively processed through the cerebral parts of our brain (those parts of our brain that provide us with consciousness and with the more sophisticated forms of reality testing). Instead, they initiate responses to situations that are relevant to something that occurred historically and, unfortunately for us, these unconscious reactions often have little relevance to our present circumstances. The mechanism that is responsible for amygdala scripts, once useful for our survival, now hinders us.

Still today, when we find ourselves in a situation that is similar to one that has historically been significantly painful for us, instead of analyzing it, our bodies react as if we are in danger, and the conscious analytic part of our brain tends to be drafted and lead us to react as if this danger is real. Survival trumps reality. Daniel Goleman, in his book *Emotional Intelligence*, describes the amygdala-mediated emotional responses as "hijacking" the cerebral cortex.[9] The primacy of the emotional response over the more analytical processes in the brain is further explained by the evolutionary history of the neocortex part of the brain. The neocortex was a late addition to the brain cavity; an add-on option that evolution offered to only a few species.

Evolution's paramount concern is survival and propagation, and the chances for survival in the face of the many threats that our ancestors faced would have been significantly diminished were it not for their amygdalae. For many millions of years, the amygdala had effectively served the survival efforts of many species, and it had done so without the more cumbersome intervention of consciousness (that is, without the intervention of the neocortex). As the neocortex evolved, the amygdala remained, in important ways, aloof from this newcomer to the mammalian brain cavity. We might imagine that the amygdala, with its focus on survival, was rather unimpressed with the realism and consciousness afforded by the neocortex, and so the amygdala did not bother to develop a lot of collaborative interaction with that part of the brain that is dedicated to higher reasoning.

Most of us closely identify with the neocortical parts of our brains. As a result, we are surprised, confused, or worse yet, self-critical when intense, nonrational thoughts arise from sources for which we have no ready explanation. From a scientific standpoint, it is not at all surprising that when an amygdala memory is activated by a cue, it circumvents the more conscious aspects of our brains and directly initiates an emotional response in our bodies. This prepares us, immediately and automatically, for flight, freeze, or attack behavior.

In order for the amygdala memory function to be an effective survival tool, it has two additional and useful characteristics from the standpoint of the survival of a species. One is that painful memories of threatening situations do not easily fade away; quite the contrary, these memories are famously robust.[10] Once learned, a bad experience that was either repeatedly painful or traumatically painful can be "remembered" for the rest of one's life. It is best from a survival standpoint to respond strongly and protectively to future similar situations, even if those situations occur years later.

The second characteristic that aids in survival is that any situation that is fairly similar to the original situation can act as a cue to activate the emotion memory. The emotions that are activated prime our body for flight, freezing, or aggression in the face of a perceived threat. If we are to survive as a species, these responses are very useful, and so the emotion memory is easily activated. From the standpoint of survival, it is much better to have a lot of false positives (falsely assuming that we are in danger of being hurt when we are not) as compared to having even a few false negatives (situations in which we *are* in danger but neglect to react in a self-protective manner). From the standpoint of a satisfying daily experience uncompromised by inappropriate emotional responses, the opposite is true.

The bottom line of evolution's development of brain physiology is that if we have been traumatized or even just significantly hurt (say, by witnessing a tiger attack, or by being shamed, painfully rejected, physically abused, and so on), then later situations that have some similarity to the original situation will activate us again and again throughout our lives.[11] Evolution was not overly concerned if a few traumatized people became impaired by their overreactivity. Neither was evolution markedly swayed if even a much larger number of individuals were only partially impaired by their reactivity; usually their partial impairment did not substantially affect their survival efforts in getting food, finding shelter, and mating.

Today, we expect of ourselves a level of psychological sophistication that is unprecedented in the vast history of human existence. We want to interact at a high level of interpersonal skillfulness with those we meet. We want to have truly intimate loving relationships with our partners and our friends; we want relationships that are not limited by scripts that get activated. We consider fearful, angry, or sad feelings that do not fit our immediate circumstances as a problem, and we want to feel generally good about ourselves and others. It is probably good that we have these high expectations, although they cause us difficulties. They point us toward a level of psychological development that is potentially very beneficial. However, few people get through childhood, never mind through life, without experiencing the kinds of painful events that create amygdala scripts, and scripts clearly act as obstacles to the level of psychological well-being that we strive for. It behooves us to relax and assume an attitude of patient generosity toward ourselves.

Psychological Problems: They're Looking a Lot Like Amygdala Scripts

Before we move on to a deeper understanding of how we can master scripts, it will be helpful to summarize the characteristics of amygdala-mediated scripts and the purpose they have served in our evolution. I have included a few citations for some of the research that supports each of the propositions below. A more complete description of much of the relevant research is summarized in appendix D.

One of the defining characteristics of amygdala scripts is their emotional nature. We have noted that when an amygdala-mediated memory is activated it includes the release of hormones into our body, creating emotions.[12] We have also pointed out that these emotions exist for a reason: they have served evolutionary survival by promoting fight, flight, and freeze behaviors.

Another aspect of amygdala scripts that we have discussed is their unconscious nature. For example, when an amygdala script is activated we do not consciously realize that we are being influenced by an emotion memory. As mentioned in the introduction, the hormonal activity that occurs with an amygdala-mediated response selectively inhibits parts of the brain central to conscious awareness—areas of the prefrontal cortex.[13] The activation of an amygdala script circumvents much of the usual reality testing that occurs in this part of the neocortex. The result is that amygdala scripts affect how we feel and how we interpret our immediate circumstances, while offering limited opportunity for conscious awareness and conscious moderation of our reaction.

Other aspects of amygdala scripts (which we have alluded to) are that they are created by painful experiences,[14] and exist in the long term.[15] In addition, they are easily activated.[16] In table 2.1, I have listed these characteristics of amygdala scripts along with their usefulness in evolutionary survival. If we compare the characteristics of amygdala scripts described in the first column of the table to what psychologists have long noted about psychological problems, the similarities become striking.

For example, beginning with Freud, the underlying basis for psychological problems has been considered to be unconscious. Amygdala script theory underlines the unconscious nature of the origins of psychological problems but requires a much less complicated model of what constitutes the unconscious than that proposed by Freud. While Freud's view of an unconscious filled with early primitive memories is not proven wrong, amygdala script theory suggests that most psychological problems can be accounted for by the unconscious recording and activations of memories that later affect a person's immediate experience. From this perspective, the unconscious that is relevant to psychological mastery is not a dark, mysterious place filled with dangerous impulses. The unconscious, instead, simply consists of emotion memories or amygdala scripts—phenomena that, as we have seen, are well documented in the brain sciences. This formulation turns out to be similar to the notion of the unconscious that is implied in cognitive therapy; treatment begins by teaching clients how to become conscious of the costly dysfunctional assumptions that underlie their problems.[17]

The concept of an emotion component of amygdala scripts also coincides with our knowledge of the characteristics of psychological problems. Negative emotions have long been considered inherent to

TABLE 2.1

Characteristics of Amygdala-Mediated Memories

Amygdala Scripts	Brain Correlates	Probable Original Function
The memories stimulated by the amygdala are unconscious.	When a script is activated, the assessment and evaluation processes that occur in the neocortical regions of the brain are largely bypassed and inhibited.	Not limited by the more ponderous analytic processing of the neocortex, we are readied for quick, unhesitating, defensive or aggressive actions in a threatening situation.
When activated, amygdala scripts include an emotional reaction.	The limbic system secretes hormones causing physical sensations that are experienced as "emotions."	The emotions are uncomfortable. They motivate protective aggressiveness, flight, and/or freeze responses.
Amygdala scripts are the result of traumatic or painful incidents.	The more painful the stimuli the more likely the amygdala will record it.	By recording painful events as scripts that can be enacted when a similar event might occur, survival is better assured.
Amygdala scripts are robust (not easily changed).	The amygdala mediates the storage of robust emotional scripts.	It is better to "remember" unusually painful stimuli, even for a lifetime, than to forget it and miss ominous cues.
Amygdala scripts are easily activated.	Amygdala-mediated memories have been shown to be easily and repeatedly activated with little deterioration under normal circumstances.	Better to be safe than sorry. From the standpoint of survival, this is important even if it results in a lot of unnecessary defensive reactions.

psychological problems. Insight-oriented therapists generally regard emotions as creating the basis for dysfunctional thinking and behavior. Cognitive-therapy theory acknowledges the important role that painful emotions play in psychological difficulties but reverses the cause, suggesting that although negative emotions (or the avoidance of such emotions) might typically bring someone into therapy, dysfunctional beliefs

are the cause rather than the result of these bad feelings.[18] The amygdala
scripts model suggests both are true: emotions and dysfunctional cognitions coexist in an amygdala script and are mutually reinforcing.

Amygdala script theory and clinical wisdom also share the assumption that early hurtful events are the precursors to later psychological
problems. Freud was the first researcher to convincingly connect
painful events in childhood to maladies of the mind. In the 1970s and
thereafter, when psychology developed refined research tools, this
proposition was repeatedly tested scientifically, with the result that the
relationship between previous painful events and most psychological
problems has now been well established.[19] (A notable exception to this
would be more extreme forms such as schizophrenia and some of the
bipolar disorders.) When we list the characteristics of amygdala scripts
alongside those of psychological problems, the similarities are striking—we are apparently looking at one phenomenon from two different
perspectives.

Table 2.2 compares what we know about amygdala scripts with characteristics common to most psychological problems. When we compare
what researchers in brain physiology have discovered about amygdala-
mediated memories with the distinctive features of most psychological
problems, we are compelled to assume that no other mechanism so readily explains what underlies much of what we consider to be psychopathology. It seems likely that while much is yet to be discovered, the
amygdala and its moderator, the prefrontal cortex, can expect to continue
to predominate as primary players in the brain science of psychological
problems. A further discussion of the role of the amygdala in a wide
range of psychological problems can be found in appendix B, "Neurobiological Research Notes."

At this point we have explored why amygdala scripts exist and
learned something about how they operate. We now turn to a more
detailed discussion of what constitutes a script and an overview of how
to learn to master them.

Facing a Script:
The Three Components of an Amygdala Script

Below are some examples that help us take a closer look at how the three
components of an amygdala script operate in everyday life.

TABLE 2.2

Comparison of Amygdala Memories and Psychological Problems

Amygdala Scripts	Psychological Problems
The memories stimulated by the amygdala are unconscious.	Psychotherapy practice assumes that most people lack conscious insight into what is bothering them.* For example, cognitive therapies help clients discover dysfunctional thoughts and assumptions that have been operating outside of their conscious awareness.
When activated, amygdala scripts include an emotional reaction.	Hurtful emotions—whether these are the pain that brings one into therapy or the underlying contributors to dysfunctional behavior—are generally understood to be a central aspect of psychological problems.
Amygdala scripts are the result of traumatic or painful incidents.	Although for the more extreme types of psychological problems the role that previous events plays is unclear, the relationship between previous painful events and the vast majority of psychological problems is well established.
Amygdala scripts are robust (not easily changed).	So are psychological problems. There would be no reason to read this book if changing dysfunctional psychological reactions was easy.
Amygdala scripts are easily activated.	Once established, anxiety, depression, and most other problematic psychological difficulties tend to be easily activated.

*See, for example, J. B. Ruskowski and D. Miller, "Shifting Modalities of Communication in Psychotherapy: Use of the Lowenfeld Mosaic Technique to Promote Self-understanding and Mental Health," *Journal of Projective Psychology and Mental Health* 13 (2006): 55–60.

Tom's parents believed in a "spare the rod, spoil the child" philosophy of child raising. He came into therapy because of his own aggressive tendencies. In therapy he discovered that he became very angry whenever he felt threatened, particularly when it seemed as if he might be found to be in the wrong.

When Millie was twelve years old, she was raped. From that day on she often felt ashamed and frightened around people she did not know. She became shy and withdrawn.

As a child, Jim was often made to feel stupid by his father. As an adult he would avoid offering opinions, and often when significant decisions had to be made he felt fearful that he would make a huge mistake.

Joan's parents were rarely appreciative and complimentary toward her. She remembers her growing-up years as sad and lonely. It was only when she did something especially well—like get straight As on her report card—that she could expect to be praised and to receive a modicum of loving attention from her family. As an adult, she prioritized work and accomplishments over relationships and fun, leading a tense and driven life. If she had a setback at work, she would feel unduly anxious and depressed.

Arthur grew up with two older brothers who took out their frustrations on him. They often ridiculed and teased him. As an adult around groups of men at work or in social situations, he always felt one down—as if he could not measure up. Even when people seemed to like him, he assumed that eventually they would reject him or otherwise hurt his feelings. He protected himself by staying peripheral in most groups and avoiding many social situations.

These people have much in common. They all have a history of either repeated painful experiences or a significant traumatic experience or both. They all have a residue from those experiences that limit their abilities to enjoy their lives—an amygdala script. In their daily lives, they each encounter various situations that trigger their script. When they first decided to work on their scripts, none of them understood what was being triggered. They simply felt that they were somehow "messed up." A conscious awareness of their script remained unavailable to them. Also, they all had developed mostly unconscious strategies to try to avoid the activation of those old painful feelings.

It is helpful for parents and other child care people to know that everyone inevitably develops amygdala scripts. Evolution has seen to it that in every human brain there exists the mechanism for this to occur. Parents, no matter how capable and loving, will not be able to protect their children from painful events that leave the indelible mark of an emotion memory, an amygdala script. We can guess that unusually sensitive children—those whose abilities for compassion and empathy will probably surpass that of their peers—will, in addition to their special strengths, be especially prone to the development of scripts. As parents, teachers, and child care specialists, the more we are aware of this mechanism, the more able we will be to help our children identify, befriend, and master their scripts. A knowledge of amygdala script theory also aids us in quickly recognizing those situations that might produce a problematic amygdala script and intervene skillfully to allay or soften the creation of a script. I have included an appendix for those whose responsibilities include the care of children: "Amygdala Scripts and Child Care."

Each of the people above has a story that depicts the three components of an amygdala script. When Tom experiences confrontations or criticisms or even disagreements, a fearful and angry feeling arises in his body (the emotion component), and at these times he says: "It feels like I'm going to get squashed. I don't mean just squashed emotionally, it actually feels like I'm going to get physically hurt [the image component]. Even though I know this isn't true. I've discovered that when I feel this way, I fear that I am not powerful enough to defend myself. I feel weak [assumption about himself] so I overreact to protect myself. It's as if my anger is a reassuring statement to myself and others: 'You better not mess with me. I'm really quite powerful.'"

Millie, the woman who had been raped as a twelve-year-old, stated that in many social situations she felt fearful and dead inside (feeling component). With a little practice she realized that in those situations: "It seemed like others, especially men, wanted to make me feel bad about myself and no one would actually be interested in me in any way that was truly appreciative of who I am [image component]. Also, when this is activated, I feel like I am somehow damaged goods, dirty, and unacceptable [belief component]. I can now see that this is how I felt right after the rape. Like everyone could see that I was damaged and no good."

As a child, Jim was made to feel stupid by his father. "When I'm at work I bend over backward to not make mistakes, checking things and rechecking them. I only take on tasks that I am confident that I can do fairly easily and I get quite fearful [feeling component] if I'm asked to do something new or challenging. It's like I think that at almost any time I'm in danger of getting into trouble [image component]. If I make a mistake, I feel like the consequences are going to be humiliating [image component and some feeling component]. At these times I can't shake the thought that I'm not really capable of succeeding [belief component]. It seems like at any moment I'm going to be found out to be a fraud [belief component]."

Joan's parents remained emotionally aloof and were often critical. The limited praise and appreciation that she did receive almost always occurred when she had done something especially well. As an adult, she calls herself a workaholic: "It seems like my success as an executive, as an athlete, and in terms of how attractive I make myself is the basis for how valuable I am [belief component]. I imagine that people are only interested in me in regard to those things [image component]. When I don't do well or even if I do just an average job, I see myself as a failure [more belief component]. When I think even of the possibility of screwing something up, I feel sick to my stomach and panicky [feeling component] Obviously, I don't think I have much to offer apart from my accomplishments. That is the only way in which I feel I can be important and loveable [belief component]."

TABLE 2.3

Three Components of Various Scripts

	Emotion	Image (What seems to be true about a given situation)	Belief (Buried in the image: what seems to be true about one's self)
Tom	Fear and anger	Seems like he is going to get squashed or harmed.	Weak, especially vulnerable to harm from others.
Millie	Fearful and dead inside	People only want to take advantage of her; no one truly interested in her.	Damaged, dirty, and unacceptable.
Jim	Fearful	On the verge of making a mistake that will cause him to be humiliated.	Incapable, a screw-up.
Joan	Sick feeling in stomach, panicky	People only value her for what she accomplishes.	Not fundamentally likable, loveable, or worthy of appreciation.
Arthur	Sad, lonely	People not really interested in him; an object of pity. His fate is to be peripheral to whatever is going on.	Unattractive, uninteresting.

Arthur's brothers teased him mercilessly. "When I'm in social groups I always perceive myself to be an outsider. It seems like others, for some mysterious reason, will attract people but I will always be on the outside [image component]. Forget about women—the only ones who are attracted to me are the ones who pity me and want to save me [image component]. As I watch other people interact, say on the street even, I often feel sad and lonely and hopeless [feeling component]. I think of myself as a loser. As if my brothers were right. As if there is just something about me that makes people want to reject me and marginalize me [belief component]."

In table 2.3, we see the three components for the scripts described above. Each of the people in our examples worked on their scripts for two or three weeks before they were able to identify the components listed below. It is primarily when a script is activated that it influences how these individuals feel, image their surroundings, and think of themselves. At other times, when the script is inactive, they have experiences quite different from those described above and listed in the table.

In this chapter and the preceding, we have explored why the brain produces amygdala scripts, what their components are, and how they function in our daily lives. Until now, we have only been able to touch briefly upon how to master them. The next chapter is an overview of the three steps to amygdala script mastery, and the chapters following that offer a detailed description of each of the three steps.

CHAPTER 3

Overview of the Three-Step Practice for Mastering Amygdala Scripts

It has been about a hundred years since Freud introduced the concepts of personal exploration and psychological change to those of us in the Western world. Since then, psychologists have continually researched and refined various approaches to psychotherapy. It is interesting that long before psychotherapists knew about how the amygdala stores emotional memories and images, psychotherapy tools had been developed that now seem tailor-made for working with each of the three amygdala-script components. In particular, three of the mainstay tools for psychotherapy have been, and still are, mindfulness, insight, and cognitive change.

Mindfulness

Mindfulness practices were adapted into Western psychotherapy from Buddhism, and we have noted how they can be useful in helping us become aware of our scripts. Mindfulness also attunes us to our emotions. In fact, its embrace by Western psychotherapists and researchers is an outgrowth of a long-standing psychotherapeutic tradition of teaching people to identify the emotions that underlie their difficulties. Mindfulness cultivates conscious awareness of feelings such as fearfulness, sadness, anger, disgust, and so on, that, when operating in the background, set the stage for our problematic reactions. The benefits of such awareness have been repeatedly demonstrated.[1] In learning to master a script, simple mindfulness practices (which will be described in more detail in chapter 4) create attunement to our bodies, which then allows us to notice when a script is activated—we literally feel its presence—and say, "Hmm, something is getting activated here." This form of mindfulness also reminds us that what "feels true" may not fit what is going on right now. Distorted by a script, our experience may have more to do with what was going on when we learned that script than with what is going on in the moment. Mindfulness addresses the feeling component of a script.

____Insight

Whereas the mindfulness practices described above attune us to the emotions that form the emotion component of a script, the backdrop for our experience in the present moment, historical insight, reminds us of the circumstances in which we learned the script. In this context, insight means recognizing the feelings and behaviors that now haunt us as being relevant to the time in our lives when the script was created. Historical insight is the "Ah ha!" psychotherapeutic experience that has so often been highlighted in movies and television and that still often takes center stage in media descriptions of psychological change.

We have already seen how Mary used insight to gain further mastery of her script. However, we should be careful to avoid a common misunderstanding: it is popular to assume that after an important insight is uncovered, psychological change will follow automatically. This is an example of our overvaluing the power of the rational part of our mind. Insight can play an important role in the mastery of psychological problems, but it must be applied properly in order to be effective. With this qualifier in mind, we can state that the particular type of insight we are describing, when carefully applied, is very helpful.[2] By first discovering the historical origins of a script and then practicing that insight—repeatedly calling the script up and reminding oneself of its origins—a person becomes adept at separating the imagery of an active script from the reality of the present situation.

Insight also offers reassurance and validation—we realize that the emotions and especially the images that arise when a script is activated make sense when we recognize that what we are experiencing is most properly understood to be a memory. By practicing the insight tools described in chapter 5, the awareness brought about by mindfulness ("Hmm, some particular old familiar feeling is getting activated here") becomes complemented with additional important and useful information ("Hmm, something is getting activated here. It may not entirely fit the situation I'm in. And I actually know and understand which particular memory is getting activated"). This is how the specific type of insight that we use addresses the image component of a script.

Belief Change

In psychotherapy parlance, belief change is sometimes referred to as "replacing dysfunctional cognitive assumptions." What this means is that individuals' thoughts can include assumptions that distort their understanding of their present situation and mislead them into feeling and behaving in ways that are detrimental. When people identify and replace dysfunctional beliefs with more accurate and helpful beliefs, their lives become more satisfying. We have learned that embedded in the problematic feelings and imagery of an amygdala script is a belief—most often about one's self. This belief can become a lynchpin for psychological change. Cognitive therapy provides the means to change those beliefs that still "feel" true when a script is activated but that were actually mislearned when the script was first created. This is extremely helpful, and in the Three-Step Practice it is simply and powerfully applied.

In changing the belief component of a script, the statement to oneself connotes further understanding, sounding perhaps like this: "Something is getting activated here. It doesn't entirely fit the situation I'm in. I actually know and understand where I learned this particular memory that is being activated. *And I also can now remind myself that . . . [a true and functional belief about oneself that contradicts the old dysfunctional belief is inserted here]."* The first two sentences of the above statement are the outcome of the first two steps (mindfulness and insight). They increase mastery of a script while also paving the way for the third step: the application of simple cognitive strategies that have been shown to be especially effective in changing habitual (unconscious), dysfunctional thought patterns.

The following chapters provide detail as to the *how* and a bit more about the *why* of these three steps: mindfulness, insight, and cognitive change. For most people, it is surprisingly easy to learn and practice these steps. In preparing yourself for that task, a few things should be kept in mind. One important point to remember is that the mind is not a stationary thing. If we want to change it, we have to engage it as a process—

working with it in the moment. We can tell wonderful stories about what we have learned about our mind, why it is as it is, the interesting insights that we have discovered, and so on. There is nothing wrong with that, but when it comes to actually establishing more mastery over our minds, such knowledge, unpracticed, is of limited value. Conceptual understanding takes a back seat to actually working with our minds—a back seat to actually becoming mindful of what is happening and skillfully engaging it in the moment. Our minds exist as ongoing processes, and the more we understand how we unconsciously allow ourselves to be defined by this flow of thoughts and experiences, the more able we are to take control and practice mastery of that flow.

Last, and most important, it is highly recommended to keep in mind that we are doing these exercises in order to help ourselves and in so doing to help those around us. In other words, working with our minds is based on an act of compassion, first toward ourselves and then increasingly toward others. That we are learning to be kind to ourselves is fundamental to working with our minds. Therefore, our approach to engaging our minds should be in keeping with this attitude—an attitude of kindness to ourselves. In approaching a path of psychological growth, people sometimes take a somewhat critical and overweening attitude toward themselves. Something like: "Okay, buster, it's about time that you got rid of those bad behaviors. Here's a way for you to finally get a grip and quit being such a [weakling, jerk, nerd, neurotic]." Such an attitude, even in mild forms, runs contrary to the whole purpose of mastering amygdala scripts, which is after all, inspired by a caring concern for oneself. All of the most effective healers approach their clients with acceptance and caring concern. This is especially important when you are the healer and the client is yourself.

If being kind and caring toward yourself is difficult, don't give yourself a hard time about that. It does not seem like a very good strategy to be critical of yourself for being critical of yourself! If you have a long-standing tendency to be critical of yourself about almost everything, simply relax and do the best you can in developing a more caring attitude toward yourself. Remind yourself that "doing the best that you know how under the present circumstances" is all that you can rightly expect of yourself.[3] In chapter 6, when we explore the third step, the cognitive change step, more will be said about how to cultivate a kinder and gentler attitude toward oneself, but even now, as you prepare to begin this

path, it is helpful to avoid any tendencies you may have toward an attitude of a stern task master or dictatorial parent who is going to save you
from your weaknesses. Instead, try to cultivate an attitude of caring help-
fulness toward yourself, as you would if you were approaching an injured
friend or child. Mastering our minds is based first of all upon a compassionate appreciation of what we have been through.

With this in mind, it becomes clear that the guidelines and insights
that follow are best understood as a process of joining with yourself,
noticing and working with your thoughts and feelings rather than of trying to force them to change in some kind of top-down authoritarian fashion. A good analogy that we will sometimes employ in the following
chapters is learning to ski or ride a horse. In skiing we learn to work with
gravity, and in riding a horse we learn to work with the movements of the
horse. Both of these entail a process of mastery rather than forced control. Success comes in realizing how exhilaratingly freeing it can be to
masterfully join the force of gravity as your skis move downhill, or in
joining with the powerful movements of a horse's body and the horse's
mind while skillfully guiding it. The Three-Step Practice for working
with an amygdala script can be understood to work similarly: identifying
and engaging the flow of feelings that underlie a particular amygdala
script, appreciating its nature and how it came to be, and finally, mastering it and modifying it so that it actually works with you. You are capable
of becoming free from the habitual constraints that an amygdala script
has imposed upon you. You will find that you can even develop a playful
relationship with an amygdala script. It is as if the amygdala is a head-
strong dragon that you have tamed. While you still occasionally need to
rein it in, once you have trained it you can learn to play with it and not be
embarrassed about sharing it with others.

CHAPTER 4

_____ Step 1

MINDFULNESS AND THE
EMOTION COMPONENT OF A SCRIPT

When an amygdala script has been activated, most people, at least in ret-rospect, can intuit that there was an emotional piece to their experience. We begin this chapter with a statement about emotions that may seem surprising to some people: sensations in our _bodies_ are what provide us with access to our emotional life. If as children we had been taught to be aware of subtle bodily sensations with the rigor applied to teaching us how to read or to analyze problems, this chapter might be very short. This is because our emotions actually manifest in our body, not in our brain.

When the brain activates an emotion, it is through our body that we can become aware of that emotion. The limbic system regions of our brain initiate the release of emotion-producing hormones. Those hor-mones create physical sensations in our bodies, and our brain then rec-ognizes these sensations as emotions.[1] This process seems unnecessarily complicated: the brain first activates hormones that create an emotional response in our body, and then the brain monitors our body so that it can become aware of that emotion. The convoluted nature of this process stems from a misstep of timing in our evolution. The evolution of emo-tions preceded the evolution of consciousness—animals that were our predecessors in the evolutionary process had developed the ability to have emotional reactions but not the ability to have a refined conscious aware-ness of those reactions.[2] This is because conscious awareness of emotions depended upon the development of an additional neurological mecha-nism in the human brain—specialized areas in the prefrontal cortex, or more specifically, a region of the brain called the "medial Brodman's Area 10" or mBA10 of the brain. The mBA10 is larger and more complex in humans than in any other species. It works in conjunction with another area of the human brain, the anterior insula.[3] This region constantly monitors internal states in the human. When the anterior insula detects indicators of emotions, it alerts the mBA10. Acting together, the prefrontal cortex and the anterior insula afford us the opportunity to be conscious of the activity of our limbic system that we call emotions.[4]

The fact that the brain relies on physical sensations to produce mindfulness of emotions converges with clinical practices that support the value of attending to physical sensations as a means of identifying emotions.[5] And so our journey toward understanding and working with amygdala scripts begins with a focus on our bodies. This approach may seem a bit odd for those of us in the Western world who tend to distinguish our minds from our bodies. Our separation of mind and body has led us to assume that to understand our minds we must focus on our minds, and to understand our bodies we must focus on our bodies. Thanks in part to insights provided by brain science, however, we can now state that having a greater attunement to our physical sensations promotes a greater attunement to what we have historically consider an arena of mind—our emotions. To better prepare us for practicing an increased awareness of the subtle sensations associated with our emotions, it is helpful to realize the implications of a very long history of separating mind from body.

____The Mind/Body Dilemma

When someone refers to "my mind," they are usually referring to an internal reference point for one's self: the experiencing, thinking, emotional being that characterizes who we think we are. We have a variety of characteristics that we might use to describe that self, and these characteristics are usually not materialistic, such as "I am smart or stupid or funny or boring or serious or silly or shy or egotistical." The self is defined in contrast to the non-self or that which is separate from me—the external world, "out there." [6] Our mind is generally assumed to be the home base—the nonmaterial "I" who ventures into and interacts with the material external world. For example, as I write this I am eating a very flavorful apple. I think of the apple as an object that is material and external to me. I consider the flavor to be an aspect of the apple (external and material) that is experienced by me (internal and nonmaterial). The "I" who experiences the apple is a nonmaterial, internal entity that I choose to call my self. My mind encompasses my self. I experience my mind as internal and nonmaterial.

I go through my daily life acting as if I am quite clear about that which is my self (my mind) and that which is other (external and material). Yet, when I consider the status of my body, the boundaries become blurred. My body is clearly made up of organs and cells; things that are certainly of the material and external world. "I" can observe my body. "I" can feel my body. "I" can do all sorts of things with my body. So I readily see my body as separate from "me." However, it is also apparent that without my body, "I" seem to no longer exist. Casting further doubt on the independent existence of my self is the fact that it is through the sensory apparatus of my body that I experience all of the external world. I cannot go anywhere or experience anything without the cooperation of my body. I experience my internal, nonmaterial self as separate from the material, external world but I also experience my material body as inseparable from my self. Is my body separate from my self (my mind) or are they inseparable?

_____1,600 Years of Mind/Body Separation

This is an age-old dilemma: one that the Western world has historically resolved by consistently taking the stance that our minds and bodies are decidedly separate.[7] If we look for a time in Western history when body and mind were not so strongly separated, we have to reach back to the pre-Christian pagan era. The pagans, whose beliefs held sway in Europe for many millennia, celebrated mind and spirit through special dance forms and rituals that suggested a unity rather than a separation of body and mind. What we know of these cultures suggests that the joy and unabashed appreciation they held for things physical was on par with those things mental.

It was in the fourth century B.C.E. that both Plato and Aristotle sowed what were perhaps the first influential mind/body separatist seeds, questioning the basis for what had previously been an easy, undifferentiated relationship between mind and body. For slightly different reasons, both Plato and Aristotle suggested that perhaps there are domains of the mind that exist separately from the domains of the body. However, the Platonic/Aristotelian concept of mind and body allowed for grey areas of overlap.

This changed in the fifth century, when St. Augustine suggested a much more radical mind/body split. He propounded that "man" is made of two "substances," one being immutable Truth that is God and the other being the much inferior body. During later medieval times, his schism between body and mind became solidified into religious dogma—body became associated with the impure and sinful, while mind in contrast existed in proximity to the soul and to God.

In the eighteenth century, "The Age of Reason," Western philosophy questioned blind adherence to religious dogma and attempted to establish a metaphysics based purely upon reason and experimentation. This offered an opportunity to revise the concept of a mind/body schism. René Descartes embraced this task, bringing the issue of the mind/body split to the forefront of philosophy. In the end however, Descartes did little to change the previous assumption. In his Sixth Meditation, he strongly separated the domains of body and mind—a distinction popularized as "Cartesian dualism"—and this assumption settled the matter for most of Western philosophy up until more recent times.

Given our 1,600-year-long history of marginalizing the body's connection to the mind, it is no wonder that today most of us still view mind and body as having distinct and separate domains, with the mind having an elevated status and the body a lowered one. For hundreds of generations, largely under the auspices of the church, Augustine's belief that our bodies are separate from our minds and the body is inferior to the mind aspect of ourselves has been woven into the fabric that underlies our everyday assumptions about reality.

It is tempting to assume that modern thinking has moved us beyond Augustine's fifth-century dogma. Most of us are, after all, willing to accept a less prejudicial view of our bodies than those who preceded us even a generation or two ago. We no longer blush and change the subject when sex, menstruation, masturbation, genitalia, and a host of other previously verboten subjects are mentioned. We have recently come to be more accepting of not covering up our bodies, at least not so completely as was required a few generations ago, and we have taken keen interest in exercise, yoga, and other activities through which we show an appreciation for our bodies. Furthermore, in modern philosophy it is popular to discount the mind/body split.[8] Nevertheless, in spite of our impressive knowledge of interactions between mind and body and our growing appreciation and acceptance of our bodies, we still maintain a subtle adherence to medieval assumptions that pretend to markedly separate our minds and selves from our bodies.

For example, when people are asked to describe themselves, they will probably come up with attributes that form the basis of their self-identity: they might identify themselves with their work, or their accomplishments, or some political or sexual preference that differs from the predominant culture, and so on. It is the rare person who would describe him or herself as a physical body moving around in the world, a body that is experiencing a continuous flow of perceptions, sensations, feelings, and thoughts.

The problem with this is not that a distinction between mind and body is no longer useful. Our consciousness in fact allows us to observe aspects of our experience that we associate more with our bodies (for example, the pleasant sensation of a warm bath) and those that we associate more with our minds (for example, an abstract concept). These distinctions remain useful and informative. What has become costly and confusing, however, is the degree to which we have come to habitually ignore our bodies as a valuable resource for tuning into our psychological experiences.

By tuning into our body, we can become more aware of that which manifests in it—our mind. The physical sensations that are our emotions hold promise for the ready access to important aspects of our minds. Although Freud had great respect for the power of emotions, he did not realize the degree to which psychological well-being might be enhanced through greater attunement to physical sensations. It is indicative of how strong our alienation from our bodies has been when we consider that we have been doing psychotherapy for over one hundred years, while mostly ignoring the very sensitive instruments that are constantly recording and signaling psychological processes, instruments readily available to us—our bodies.[9] Recently it has become obvious to a growing number of psychotherapists that physical awareness is a 'via regia', a royal road, to identifying various emotional (psychological) states.[10]

Most of us have lost much of the awareness of our bodies that characterizes a child's relationship with his or her body. We have given up the very rich and intimate communion that we can have with our bodies. This inattention is so commonly accepted that we do not view it as much of a problem. We do not notice that our conversations, especially our most intimate communications, are inspired by sensations that arise in our bodies. We joke, and our humor flows through our bodies, expressing itself in physical laughter. We vigorously debate each other because in our heart area we feel an urgency to be acknowledged or because in our stomach area we feel a fearfulness that something important will be missed. We sigh, feeling in our bodies the disappointment when another person does not understand us, or we feel irritation energizing our bodies when we are discounted, feel attacked, or are threatened by someone's criticism.

All of these bodily sensations represent information—important and potentially very useful information—that is usually lost to our awareness. Most of our emotional information is ignored because we habitually believe that what is really happening is the removed, disembodied abstractions that occur in the very small area of our bodies called the cerebral cortex—the area of the brain through which we label, interpret, and explain. At least in much of our conscious awareness, the richness of experience and information provided by the vaster reaches of our bodies' sensations is lost to us.

Approximately one hundred years ago, Freud discovered the prominent role that identifying and engaging emotions can play in a psychotherapeutic process—one of the three major insights that he presented to the Western world.[11] Since Freud, the role that emotions play in psychological well-being has been delineated and refined. In retrospect, Freud's insights into the impact of emotions on our day-to-day existence almost inevitably led to a recent innovation in Western psychology—adoption of the Eastern practices of mindfulness.

Mindfulness Practices

We have already noted that the term *mindfulness* and the practices that underlie it are borrowed from Buddhism a religion rooted in exploring the mind.[12] Buddhist mindfulness has three aspects—awareness of body, awareness of speech, and awareness of mind—but the particular form of mindfulness that concerns us here is usually associated with mindfulness of body. Simple "mindfulness of body" practices have been engaged in for thousands of years as a means to provide a bridge that heals the gulf between body and mind.

Once while walking with a long-time Buddhist practitioner, I asked her to describe mindfulness of body. She began by describing in detail how as we walk, we can feel our legs swinging with each step and feel the contact of our feet on the ground. She noted that at the same time we can notice the emotions and sensations that arise in our body, especially our torso: for example, the energetic heart-pounding feeling when we see a puppy dog wander into traffic on the road; the passing sexual feelings that arise and suffuse through our bodies as an attractive person smiles flirtatiously; the warm feelings that arise in our heart area as we note being together on such a nice day; the sad feeling in our heart that arises when we mention a companion who has recently died; the mild sick feeling in our stomach when we remember acting foolishly. I realized that she could go on in this vein indefinitely, describing an array of experiences that constantly arise in our bodies and inform us.

Western Psychology's Assimilation of Mindfulness

In Western psychology, good clinical practice awaits good research before it embraces new tools and perspectives. It is only within the last decade or two that clinical psychology has developed strong empirical research demonstrating the effectiveness of mindfulness.[13]

Mindfulness was originally introduced into Western psychotherapy largely due to the research of J. Kabat-Zinn, Marsha Linehan, and more recently, Segal, Williams, and Teasdale.[14] Kabat-Zinn showed mindfulness to be a useful tool in the treatment of chronic pain, stress, and anxiety.[15] However, Linehan was the one who brought this tool into much wider acceptance when she developed an effective therapy strategy for patients who have historically been famously difficult to treat. [16] In her treatment protocol, mindfulness played a central and effective role. Mindfulness practices were effectively utilized by Linehan to teach people awareness of their bodies as a means of becoming more aware of their emotions and so pave the way for them to manage their emotions better.[17]

In a separate line of research published in 2001, Francine Shapiro developed a psychotherapeutic tool that also relies upon teaching clients to be mindful of their emotions. Her approach, like Linehan's, has been strongly supported by independent empirical research.[18] Although Shapiro does not use the term "mindfulness," she teaches people to focus on subtle physical sensations associated with their emotions. Her approach is especially interesting in that it includes the means to focus directly on the particular emotion that underlies a particular psychological problem.[19]

Because Shapiro's original work was mostly with people who had been traumatized, the emotions that were associated with their problems were quite intense. The intensity made it relatively easy for Shapiro to help her clients associate those emotions with various areas of their bodies. Once clients were able to associate painful emotions with a particular area in the body, Shapiro would ask them to describe the physical sensations as specifically as possible. This mindfulness procedure represented the beginning steps in a very effective therapeutic procedure commonly referred to as EMDR (Eye Movement Desensitization and Reprocessing). EMDR is now being adapted and applied to other types of psychological problems as well.[20]

Edna Foa has suggested the usefulness of another application of mindfulness of physical sensations. She has shown that conscious attunement to physiological sensations increases the effectiveness of exposure treatments.[21] Various researchers have shown the utility of being mindful of bodily experiences as a very useful tool in working with anxiety.[22] Other researchers have documented in general the positive effects of mindfulness.[23] To create the first step of the Three-Step Practice, we have borrowed from those discoveries of Buddhism, and from the Western researchers mentioned above, to create a simple and direct application of mindfulness.

____Engaging the Mind as a Process

In preparing for step 1, it is important to remember that what we are doing is engaging a process. If you practice the mindfulness exercises described in this chapter, you will discover that your mind is very dynamic. In fact, the mind is not so much a "thing" as it is a process. If we carefully watch our mind in action, we will see that what we actually experience through our minds is a continuous, ongoing, changing, moment-to-moment, active process that shapes our internal and external realities. Nothing that we sense, deduce, intuit, or figure out—nothing about existence and the way we experience it—is based upon a static experience. All that we know is the result of processes that are starting up, manifesting for a short while, ending, and then reoccurring in a different form.

Freud's view of mind emphasized structural mechanisms rather than dynamic processes. He conceived of the mind as made up of the id, ego, and superego—three components that interact in fairly simple patterns. He did acknowledge the dynamic aspect of our mind through his use of "free association," by which patients were asked to access the stream of their thoughts and emotions by stating everything that came into their consciousness. Freud's purpose in doing this, however, was to offer the

therapist ("analyst") an opportunity to extract from that stream important unconscious information and to then present it to the patient.
Although it is clear that Freud's process of free association and interpretation led to new insights, it remains less clear how these insights could
be most effectively used to modify everyday, ongoing currents of thought
and experience.

Given our present-day neurobiological understanding, we can guess
that therapy will be most efficient when it directly addresses the processes
of the mind while they are occurring—a "process-oriented" therapy.
Although generally unacknowledged, just such a process-oriented view of
the mind is fundamental to the tools of modern cognitive-behavioral psychology—tools that for the past fifteen years have dominated the world
of proven psychotherapeutic approaches. These approaches have as their
basis seeing the mind as a series of processes that clients are taught to
identify while they are occurring, particularly those processes of thought
and behavior that underlie psychological difficulties. After identifying a
habitual thought/feeling process, clients can then be shown how to apply
various strategies to change those processes. This, however, is not how
cognitive therapists understand their work. They have been slow to
appreciate and capitalize on the power of working with an activated mental process. Instead they tend to adhere to the Freudian strategy whereby
something important (for cognitive therapists, a belief; for Freud a historical insight) is extracted from the stream of thoughts and then utilized
in the ensuing therapeutic process. Once identified, the belief is discussed
and various strategies are employed to change the dysfunctional beliefs
into a more functional form. The belief itself is treated separately from
the internal stream of thoughts and feelings that supported and generated it. Cognitive therapists emphasize changing beliefs as central to the
therapeutic gains that their clients experience.

Recent research suggests that healing dysfunctional thoughts can
occur without the necessity of extracting and changing beliefs.[24] When
clients are taught to identify and replace a dysfunctional belief, it
becomes necessary that they notice when that belief is actively affecting
their psychological well-being. In other words, a client has to tune into
the procession of thoughts and feelings that arise in the mind in order to
engage and modify a particular dysfunctional belief. Until recently, this
aspect of cognitive therapy has been considered to be an inconsequential
but necessary means for attaining the sought-after goal of restructuring

the effects of dysfunctional thoughts. Now there is a growing body of evidence that suggests that mindful attunement to thoughts—an increased awareness of ongoing mental processes—may, by itself, play the biggest role in bringing about the therapeutic change that results from cognitive therapy.[25] Beliefs do change as a result of mindful attunement to feelings and mental imagery, including thoughts. It is likely that a great deal of this change does not result from a conscious process of changing beliefs but is instead the result of separating oneself from those beliefs—standing back and seeing dysfunctional beliefs as mental phenomena that arise in a person's mind—rather than being passively hijacked by them and acting as if such assumptions are compelling truths about oneself and the world.

As we come to appreciate this, we realize that many of the activities that occur during almost all forms of psychotherapy can be inferred to be effective because they teach clients to consciously identify underlying dysfunctional processes that had previously been operating without conscious awareness and to actively engage these processes while they are occurring. The therapeutic approaches that have directly embraced mindfulness practices seem to optimize most efficiently the power of directly engaging the processes of the mind. Linehan points out that applying mindfulness helps us to identify our emotions and then manage them.[26] In the Three-Step Practice, we learn mastery through practice sessions where we purposely activate problematic scripts and practice the three steps with them. One of the strengths of this practice is that we soon become very familiar with the particular script that we are addressing, making it much easier for us to catch the script while it is actively affecting our daily life. This allows us to apply what we have learned in our practice in order to change an ongoing mental process.

––––––––

Tom had a vivid memory of himself as a child of grade school age, standing outside his rural home, listening to his parents yelling at each other, hearing things breaking inside, and hoping that no one was getting seriously injured.

––––––––

These experiences were dreadful for Tom. In the first sessions of therapy, Tom realized that the emotions that occurred at the time of those memories were, thirty years later, still being activated. During therapy, mindful awareness of the emotion memories from those times allowed Tom to use those memories as a frequent reference point for identifying and skillfully engaging his scripts. As he was finishing therapy, he described the difference between the activation of these events before therapy and after working on them in therapy. Before therapy, the emotions arose unconsciously, seeming "to come out of nowhere," and they felt traumatic. When something would threaten to awaken these emotions, Tom would habitually act out various avoidance behaviors. At the end of therapy, Tom was able to catch the process of his script and to change it. The script then became associated with a practical appreciation for who he was and what he had gone through.

Tom's experience typifies the transition that we often see: from a problematic, scripted emotional process that intrudes unconsciously into the stream of everyday thoughts and emotions to a masterful, conscious relationship with scripted thoughts and feelings. Although he had had "symptoms" that could be easily be fitted into a diagnostic category, it would be awkward and misleading to say that Tom had both depression and an anxiety disorder and that therapy had cured it. Much more to the point and informative would be to say that Tom became aware of an old scripted memory that arose in his mind unconsciously, and that he learned to master it.

_____ Beginning Step 1. Noticing When a Script Is Activated

How does one learn to be mindful of the particular emotions associated with a particular script? Do mindfulness practices require us to receive various teachings and spend hours meditating every week (like a Buddhist monk)? Sorry to say, most of us are not *that* interested in learning to work with our minds. Fortunately, there is a much easier approach that can be very fruitful in gaining the degree of mindfulness skills necessary to begin addressing most psychological problems.

First, we need to notice when a script is activated. As we have suggested, this can be directly accomplished through mindfulness of body. Most people habitually try to avoid such mindfulness. Indeed, the most difficult thing about tuning into old feelings is that it usually runs directly counter to a long-standing habit of avoiding those very same feelings—a long-standing habit of defensive behaviors. Nevertheless, if we apply what we know about the physical sensations of emotions and remember the importance of practicing with the emotions while they are activated, we will usually find that overcoming these old habits is not difficult.

When we notice that we are avoiding something or more directly notice that we are avoiding a bad feeling, we are at the threshold of identifying a script. Identifying a script means knowing that you are thinking or acting in a way that is contrary to your desire to be happy or fulfilled. Mindfulness helps us come face-to-face with the emotion component of a script. Most of us already have some mindful awareness along these lines. For example, we might notice unsettled sensations in our stomach when we are anxious. Or we might notice a sick feeling when our teenager is unaccounted for late at night. Or we might feel a pang of sadness in our heart when we think of a loss that we have experienced. Another common bodily feeling associated with an emotion is muscle tension, often in one's neck, jaw, or shoulders.

A series of small steps leads to the skill of working with the feeling component of a script. However, once you get a feeling for the smaller steps, you will be able to accomplish them as one step: facing and identifying a scripted feeling by tuning into it in your body. So do not get discouraged if the following practice seems laborious and complicated. Learning to drive a stick shift seems hopelessly complicated at first, but with a little practice you get a feel for it, and what at first entailed many steps soon becomes one action: "I backed the car out of the driveway." (And learning to be mindful of a script is usually much faster and easier than learning to drive a stick shift!)

If every day, at the end of the day, we were to set aside ten minutes and ask ourselves, "What situations occurred today that felt bad or felt 'off' or awkward?" we would be taking a big step in becoming more mindful of our scripts. This process requires some fearlessness and discipline. We have to be willing to set aside our comfortable habits of avoidance or of blaming others and, equally important, set aside less comfortable habits such as blaming ourselves and thus becoming depressed. We have to pointedly look for emotions in our bodies. Pursuing problematic emotions in this way proceeds with a process of becoming more and more concrete and specific about the situations that bother us, then focusing on what is being activated in those moments. A ten-minute "taking stock" session might proceed like this: We begin by asking, "At what points during the day did I feel that the problem I'm

working on may have been activated?" Here are some examples of the
sorts of things that might follow:

"Okay, the problem is that I often seem to alienate people. When did
that come up today? Well, Jane seemed a bit upset with me this
afternoon. What did I say to her? What was I feeling when I said it?"

Or:

"The problem is that I don't have enough self-confidence. Hmm . . . in
what situations does it feel like I don't have enough self-confidence?
Well, when I was interviewing today, I felt pretty low on the self-
confidence scale. What came up for me? What was I feeling then?"

Or:

"The problem is that I'm depressed. That's what my Doc said. What
were the situations that made me feel worse? Well, when I think of
myself at social gatherings, I feel icky and down on myself. I wonder
what that's about?"

Or:

"The problem seems to be that I'm anxious. When do I feel most
anxious? Well, it's when I'm out of the house. Is it every time that I
am out of the house? Yeah, pretty much. What is it about being out of
the house? How is that different than being at home? Well, when I'm
out, I'm among people. . . . Okay, when I was at the bank waiting in
line and that guy was talking to me and I thought that maybe he was
going to ask me out for a cup of coffee. I actually liked him, but
immediately I had this image come up that he would end up not
liking me. Is that always what comes up? Let me see if that is also
true of other situations where I felt anxious."

In the above examples, people took the time to sit down and look at problematic situations in their lives. They endeavored to be more and more
specific about what activated their script. They then looked to see what
feelings were being activated. These examples were of people who were
aware that they had some sort of psychological difficulty, and by focusing
on concrete situations that evoked the problematic behaviors or emotions,
they took an important step toward identifying the underlying script.

For people who have not identified a particular psychological problem, there is a more generally applicable form of tuning into what is being

activated. This focused approach, which involves scanning through your day with an eye on tuning into underlying feelings, might look like this:

"As I scan through the whole day, what do I notice? Hmm . . . as I took my shower, I remember feeling a bit unsettled like I do every workday. Well, probably, everyone does. Okay, be that as it may, it might be worthwhile to note what in particular gets activated for me as I'm readying myself for work. So, I'll note that for future reference.

"Let's see now, breakfast was good. I remember laughing with my partner. Hmm . . . something else there though. Yes, that's it! As I drove away I felt a little irritated. What was that about? Oh, yeah we were discussing our vacation and we disagreed. She always wants to fit in a visit with her alcoholic sister. What's with that? Actually, I avoided that disagreement pretty well. Okay, okay. I better also note that and see what gets activated for me there.

"Okay, now at work. It was a pretty good day at work. Hmm . . . there was that problem with Phil. What was his problem today? Maybe he doesn't think I'm holding up my end of that project. He does work harder than I do. Does he? Is *that* why he seemed upset? Being objective about it, I don't have any idea why Phil was in a bad mood today . . . more to the point, what bothered me about his gruffness? What came up for me? I'll make a note of that also."

In all of the above instances, the person was practicing noticing when an emotion came up that did not seem to fully fit the situation at the moment and at the same time noticing in very concrete ways what those situations were.

What do these practices have to do with mindfulness of body? Let us experiment. Imagine a recent situation that seems as if it may have activated a script for you. First, we will apply a "figuring out" approach, which is the one we were taught in school. It has nothing to do with mindfulness of body. Imagine the situation as a historical event, telling yourself that you are going to analyze the situation rationally and conclude through a process of logical induction and deduction what was really going on. Use only the rational figuring-out part of your mind. See what you come up with.

Now, say to yourself that you are going to examine the same situation, but that you are going to let your body and your feelings inform you about what was going on. Again call up a mental image of the situation, but this time, while walking yourself through what happened, maintain an easy awareness of your body. Don't overdo it. Simply review the situation moment-by-moment and while doing so passively notice what is arising in your body. Perhaps a tightness of some muscle group will come to your awareness, or a mild feeling of sadness in your heart, or an unsettled feeling in your stomach area. Stay tuned to that feeling and where it is in your body for a few moments, letting it inform you about how you were affected by the situation and also about the situation itself.

Notice if there is a difference in the information you garner about the situation using this mindfulness-of-body strategy as compared to the first, more cerebral, analytic strategy. If you notice a difference, you are not alone. Most people who do this experiment get much different information using the "listening with their body" approach as compared to figuring it out. They report having more of a "feeling," in general, for what happened and, in particular, of being more aware of how they were affected by the situation.

By applying mindfulness, people are more readily able to pinpoint and identify what feelings are becoming activated. The lesson is that in looking for scripts that are activated, you use an awareness of your body to call up more experiential (as compared to more cerebral) knowledge of what occurred. Cerebral knowledge, of course, has its place and value. It's just that it is fairly ineffective in helping us tune into scripts.

Some people find it very easy to notice their emotions and to sense that they are being activated. At this point, those people will be wondering what all of the fuss is about. They go through their lives conscious of their feelings. If you are one of those folks, then we are happy for you. Please be patient with the rest of us. Also, you may discover that mindfulness of body can help even you better master your feelings and so make them work for you—help you ride your feelings as compared to having them take you for a ride.

On the other hand, if this whole discussion about using feelings to tune into what is being activated sounds like Greek to you (or like Chinese, if you're Greek), and when you tried this last experiment you could not tell a difference between imagining the situation while being aware of your body versus a purely logical approach, then do not get too discouraged. You also are not alone. Some people find it very difficult to associate their emotions with subtle sensations in their body. Take heart—one of the more famous and effective psychotherapists of our time, Albert Ellis, seemed to have neither much awareness of nor use for emotions, but he nevertheless helped thousands of people, all the time minimizing emotions as part of the psychotherapeutic equation.

More important, the skill of tuning into emotions, like any other skill, can be learned. Remember that scripts have three components. If you find it difficult to identify one type of component, then you will probably find it much easier to be mindful of one, or more likely both, of the other two. All three are workable and helpful. So if you happen to be one of those people Carl Jung refers to as a "thinker" (not a "feeler"), simply do the best you can with what is described in step 1. Step 2 and step 3 will probably be easier for you.

Let us take the next step for working with the emotion component of scripts. Although it is not crucial, it is very helpful to associate a particular emotion with a particular area in your body. To accomplish this, again imagine a specific situation that evoked a problematic emotion and, after you become aware of that emotion, ask yourself, "Where in my body do I associate the sensations that arise with this emotion?" Pay special attention to your torso area. If necessary, keep referring to your memory of the situation that evoked the emotion so that you can keep the feelings fresh. (It can be hard to keep focusing on something so seemingly negative. If it seems undoable, it may at first be easier to do this with a therapist.) Perhaps you will notice that you associate the emotion with your chest, or your stomach, or both, or some other part of your body. Then ask yourself how you would describe the sensations in those areas of your body. You do not have to be particularly good at putting the subtle body sensations into words. Accurate articulation is not the goal. The goal is to identify the particular scripted emotions as they arise in your body, which will allow you to identify more easily when a script is being activated. You might think of the sensations in your body as a reference point—a physiological handle or tag that provides a shortcut to the emotional part of an amygdala script.

As people become more attuned to their emotions, they often discover that a particular script is activated much more frequently than they expected. This realization can be quite disconcerting. The flip side, however, is that the more frequently a script is activated, the more the benefit that can be gained from mastering it. In summary, we have described two steps in working with the emotion component of a script: the practice of identifying when a scripted emotion is being activated, and then noting the body sensations associated with that emotion .

_____Discriminating between Scripted and Other Emotions

The value of being conscious of emotions is greatly enhanced when we understand what a particular emotion means and how that information can be useful to us. This information is sometimes referred to as emotional insight. The insight derived from a particular emotion may not always be insight about a script. Three types of emotional insight are relevant here, and mindfulness will help us distinguish between them. Each type of emotional insight is characterized by its origins. Some emotions originate from past learning—those we refer to as the emotion component of an amygdala script. Other emotions originate in the situation itself—the emotions tell us how we were being affected in the moment when we experienced the emotions. The third category of insight actually originates with other people—we simply pick up on the emotions of others. This last type of emotional insight is the basis of empathy.

We obviously have to determine if the emotion that we have noted is one that is being activated by an old script or whether it falls into one or both of the other two categories of emotions. How do we discover the origins of a particular emotion? Or, in other words, in a given moment, how do we know whether we have now in our awareness a scripted emotion or one of the other two, or for that matter, a combination of any of the three? Happily, much can be learned by simply remembering to ask ourselves the questions "Does what I'm feeling right now fit with the situation at hand? Does it seem proportional to the situation? Do my emotions seem 'in keeping' with what is going on?"

Perhaps surprisingly, often with only minimal practice, an honest questioning of oneself in this way, while at the same time staying attuned to the feelings in one's body, usually yields a pretty clear answer. Please note the part about *while staying attuned to the feelings in one's body*. It is at this point that the above steps of noticing a particular emotion and isolating it, often by associating it with areas in your body, comes in very handy. Staying attuned to the emotion while questioning yourself about it helps enormously. (If you are one of those people who finds it difficult to associate an emotion with your body, simply stay with the emotion however you experience it.) The answers to the above questions may not be definitive: "Some of what I feel seems to fit, but other aspects of my emotions seem either out of place or disproportionate to what is actually going on." In fact, there is almost always at least some emotional reaction to a situation that fits that situation—some part of what we are experiencing that is simply situational.

Again, we should stress that this procedure is not as complicated as it may sound. To check, try an experiment. Think of any situation that you have been in recently, a situation that has some negative emotion attached to it. It could be an interaction with a family member, a co-worker, a boss, an enemy, or a friend. Imagine yourself in that situation. Now tune into how your body feels, paying special attention to your torso. Do not work too hard. Simply imagine yourself as an embodied person who can become aware of whatever reactions you might be having. (Again, some people insist that their emotions exist in their head, not their body. If that is true of you, simply remind yourself to be mindful of your emotions, wherever they are.) Now imagine the situation that you have chosen and note the emotions that arise as you imagine the situation. They might be quite subtle or more intense. If you can, note which parts of your body seem to be associated with those emotions. Then, when you feel that you have a pretty good attunement to the quality of the emotions that are arising, ask yourself "Do the emotions seem to fit the situation they arose in or does it seem like they are different from what I would expect if I looked at the situation objectively? Also, do the emotions seem to be proportionate to the situation I'm in? For example, if someone else were in this situation, would I expect that person to feel what I feel? Would I expect that person to feel it to the same degree that I feel it?" If the answer is *yes*, double-check it and make sure. Our tendency is to assume that our emotions can be accounted for by the circumstances, so you should be especially careful at this point. If the answer is still yes, then your working hypothesis can be that the emotion is a situational feeling.

(Okay, let's be honest. You may not have actually performed the experiment above. You may have instead decided to check out where all of this is going next. However, what follows is not going to make nearly as much sense if you don't actually spend a few minutes experimenting.)

If the answer is *no*—if the emotions do not seem to be both proportionate and fitting to the situation—then one of the other two possibilities is probably occurring: either an old script is being activated or you are picking up on another person's emotions, or both. You can check this out by further questioning yourself: "Do these emotions, which I see aris-ing, seem uncomfortably familiar to me? In other words, is this an emo-tional reaction that seems to stretch back into my personal history?" If the answer is *yes*, then you can be quite confident that at least part of what you are experiencing is, in fact, an old script. Through the two sets of questions, you have established that what is happening to you emo-tionally cannot entirely be accounted for as a result of the situation itself, and that your reaction is also an old familiar emotional reaction. This conclusion points you directly to the likelihood that, at least some of what is coming up is an old habitual script. With practice, the ability to discriminate when a script is being activated becomes quite easy.

If, on the other hand, the answer to the last question is "No, this does-n't seem to be a familiar and repeated kind of reaction for me," then because you have eliminated the other two possibilities, you are probably feeling the emotions of someone else in the situation that you are imag-ining. You are remembering someone else's reaction to the situation and you are feeling that person's emotion. See if what you know about the others in that situation and what you remember about the situation would suggest that you are having an empathetic response to someone else. Realizing that you are feeling someone else's emotions is clearest when you are in a situation with them, but it is also not unusual for remembered situations to include another person's emotions, and those emotions can actually predominate, coloring the remembered situation. We will discuss how to work with situational and empathic emotions in chapter 8. Our present focus is upon emotions that are generated by an amygdala script. The decision tree in figure 2 summarizes the questions above.

___How to Practice Step 1 of the Three-Step Practice

So what do we do with the emotion component of an amygdala script once we have identified it?

If we repeatedly evoke an amygdala script and then label its emotion component ("There's that emotion again"), we create a link between that amygdala-mediated emotion and the highly evolved areas of the neocor-tex associated with verbal learning and consciousness. When an amyg-dala script is activated, the cerebral cortex is not so out of the loop that it

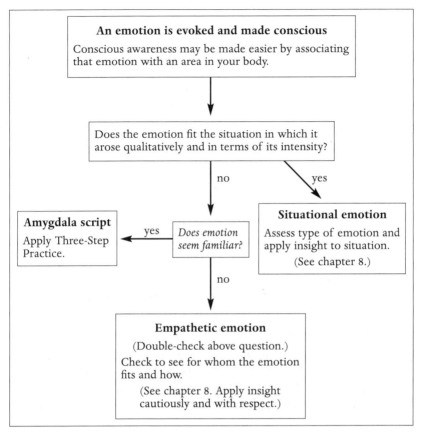

FIGURE 2

Identifying the origins of emotions

cannot function at all to mediate that script. It simply requires a bit of prodding in order to do its reality-testing job. That prodding can begin with verbally identifying the emotion component. Through repetition, we cajole the prefrontal cortex and other parts of the cerebral cortex into conditioning a new response to an old one. Therefore it is not surprising that tuning into and separating out old habitual emotions from situational emotions has long been shown in psychotherapy research to be a helpful practice.[27]

The first step toward changing an amygdala script can be accomplished by repeatedly imagining a situation that activates the emotion associated with an amygdala script, noticing where in your body you experience the sensations associated with that emotion, noting some of the particulars of that emotion, and then verbally acknowledging it: "There's that emotion." We suggest that people spend a few minutes a day imagining a recent situation that evokes a scripted emotion that they are working on, noticing it in their bodies, and then saying to themselves, "There is that emotion." (This statement suffices as a shortened form of calling upon the neocortex, of awakening your consciousness to what is going on. The long form would probably be something like "Hey, consciousness, I've got an emotion activated here and I think that I learned it some time ago. So it doesn't entirely fit the situation that I'm now in. So, consciousness, please help me not get too caught up in my old patterns of reacting to this feeling.")

Sometimes, though, the emotions are so stubbornly intense that the above practice does not decrease the intensity of the emotion. Adding another tool is then called for—a well-researched therapeutic tool called "exposure" (described in chapter 5). If the emotions that arise in doing the exercises that we have just described are too intense and unbearable, you may want to discontinue calling up the emotion component of a script until you have done the exercises in chapter 5.

Our practice should be one of working with the currents of our mind, not rowing upstream against them. What we should be attempting, *always*, when we embark on helping someone through psychotherapy or ourselves through a personal development process, is to clear away the obscurations that cloud our natural ability to enjoy our minds and our lives. The outcome of the practices that are described in this chapter, and in those that follow, are practices in seeing what is already there and in jettisoning what is unnecessary and burdensome. It is helpful to envision the exercises and practices in this book as ways of learning to let go and *not* to work so hard. In trusting your experience by paying attention to what your body is telling you, you are practicing lightening your load. It may not seem that way when you are first learning about amygdala scripts, but that is the outcome. People who properly practice what we are suggesting will find that they are cultivating an easier and more appreciative relationship with themselves. They are learning to relax and trust their minds, and especially their bodies.

This first step, along with the other two that are described in the following chapters, should be practiced "off-line." By "off-line," we mean that they should be repeated in practice sessions that are set aside from our everyday lives. How to practice the three steps together will be described in detail in chapter 7, entitled "Putting It All Together." After practicing with a script for a while, it then becomes much easier, in fact it becomes natural, quickly to apply the three steps of mastering a script when it is activated in the flow of our everyday lives.

Summary. The First Step in Mastering an Amygdala Script

1 Note what upsets you at different points of the day and pick one of these situations for practice. You may use the same situation over and over again.

2 Tune into the emotion associated with that situation, and, when possible, associate it with subtle sensations in your body.

3 Distinguish between those emotions that are old and scripted, those emotions that are situational, and those emotions that are empathetic. After we have identified a particular script's emotion component, this step can be eliminated for that script.

4 While noticing the sensations that are associated with a scripted emotion, say, "There (in my body) is that emotion."

5 Practice, practice, practice—off-line, by purposefully thinking about activating situations and repeating step 2 and step 4 above.

6 Reinforce, reinforce, reinforce. Credit yourself for your insight, your courage, and for practicing such a difficult task. Congratulate yourself after each practice session and for every bit of success that you experience. Practice and reinforcement are how behavioral changes occur.

Using mindfulness to discriminate between emotions has additional benefits that go beyond allowing us to identify and work with scripts. As we have seen, by separating scripted responses from a situation, we clarify what else is going on in that situation. In addition to identifying old scripted emotions, we are able to discern emotions that directly arise from our present situation, including emotions that we experience empathetically through others. Our discussion of emotions would be incomplete if we ignored how to use the valuable information that these other categories of emotions provide for us. Therefore, in chapter 9, we will focus upon unscripted emotions and how to enable them to work for us.

CHAPTER 5

_____Step 2

INSIGHT AND THE
IMAGE COMPONENT

A few years ago my brother was climbing a cliff in the canyon country of Utah. He had just pulled himself up over a ledge when he heard a nightmarish sound: the loud distinctive buzz of a rattlesnake right in front of him. "The next thing I knew, I was standing one ledge further down the cliff. I can't say that I looked before I jumped, either. I'm very happy to report that there was a ledge just a few feet below me." My brother had not actually seen the snake before he jumped. The sound of the rattle itself was enough to evoke immediately a vivid image of a coiled rattlesnake, at the level of his throat and about to strike. Foregoing any analysis of the situation, my brother's amygdala had signaled his body to act, and he jumped. The more conscious, rational part of my brother's brain might have preferred to take a moment to evaluate the wisdom of blindly jumping backward. Under the circumstances, the moment's hesitation that would have been required for my brother to consult with the higher-level evaluative parts of his brain might well have led to disaster.

We have seen before that when an amygdala-based memory is activated, the neocortical, consciousness-producing areas of the brain tend to be circumvented.[1] To use a modern analogy, the amygdala is like the outlaw member of an organization who "cuts through the red tape" and impulsively makes an executive decision based upon whatever information is immediately available. The information available when an amygdala script is activated is the image component—in my brother's case, an image of a rattlesnake about to strike.

It is not incidental that an amygdala script includes an image. The information in the image component of an amygdala script is crucial to its usefulness. Simply recording the emotional component of a hurtful or dangerous experience is not very helpful from the standpoint of evolutionary survival. My brother, who had a well-functioning amygdala script regarding rattlesnakes, would not have benefited much from an amygdala script that simply left him standing on the cliff frightened but with no other information. An activated feeling component of an amygdala script, by itself, divorced from any additional information, would leave its experiencer at quite a loss. Had it been the case that in the lives of our early primate ancestors scripts were activated with only the emotion component, our predecessors would have found themselves under the spell of a strong emotion—suggesting that something, perhaps something crucially important, was happening—but without the imagery necessary to decide whether to run, hide, freeze, or attack. More information than mere emotion is required in order for an amygdala script to be a useful tool for evolutionary survival. This information takes the form of an image sometimes referred to as a "gist" or an "implicit" memory.

How the Image Component Works

LeDoux is generally credited as being the first to describe the information that exists in amygdala-based memories as information that goes beyond the purely emotional.[2] This has since been demonstrated in various laboratory experiments. One example of this is found in research on conditioned responses.[3] Some conditioned responses are created by exposing an animal to a traumatic or repeatedly hurtful stimuli (such as electric shocks administered to the feet of a rat).[4] Traumatically induced "conditioned responses" are the same phenomena that we are referring to as an amygdala script.[5] In studying conditioned responses, experiments are often done where hurtful stimuli are paired with particular cues. For example, whenever a rat wanders near an area that is painted yellow, it receives a painful shock. If this has happened a number of times or if the intensity of even one shock is sufficiently high, a "conditioned response" is learned. If a maze is arranged such that this same rat comes upon a yellow area, even under very different circumstances, the rat behaves as if it is in immediate danger of being shocked, even though no shock occurs. In fact, the brain of the rat can be seen to react in a manner very similar to the way it reacted when the rat was actually hurt. In addition to the emotional response, other information is also activated—information that is intended to protect the rat from future pain. It might be that when our rat is confronted with the color yellow, it runs to an area in the cage furthest away from that color. A general idea or image of the original situation is activated and informs the animal's behavior so that it can accurately act in a way most likely to avoid the hurt. It reacts in a manner that was relevant to the original situation, even when the present situation is different and is not hurtful to the animal. In the case of the rat, the emotional component of its script is fear, but with humans the painful information recorded in the amygdala can be associated with a variety of negative emotional experiences, fear being only one of them.[6]

Humans are similarly prone to record the gist for painful emotional events.[7] When an amygdala-based memory is activated in a human, along with the emotional response, a general outline of salient information—what we are calling an image—is also activated. The image component of an activated amygdala script usually misinforms us about our present situation, giving us an exaggerated or distorted impression that we need to protect ourselves from harm.

Knowing this helps us understand our tendency to react to things in an unconstructive way during times of psychological distress (when an amygdala script has been activated). Stated slightly differently, the gist or implicit memory mechanism that researchers have identified in amygdala-mediated memories (the "image component") coincides very well with the problematic psychological distortions, which we have historically called neurotic. The image component is activated unconsciously and so unconsciously imposes itself upon the situation we are in. In my brother's case the activation of an image component of his rattlesnake script would not be construed by anyone as neurotic; this component of amygdala-based memories, however, is in most instances inappropriate to what we and others expect of ourselves.

There is another aspect of the image component of an amygdala script that is especially relevant to psychological difficulties. We have mentioned that the image component of an amygdala-based memory is sometimes referred to as an implicit memory or gist. The reason that the terms gist or implicit memory are used by researchers in describing the information that is stored via the amygdala is to contrast that type of memories with our usual idea of memories—memories that include specific detail (that is, memories that researchers refer to as "declarative memories").[8] When we recall a declarative memory, we remember what happened as something that explicitly happened. Rather than an outline or gist of what happened, we remember particular details, such things as place, time, events, and participants. Typically, these details arise in the form of a visual memory of what has happened.

In contrast, an implicit memory or gist memory suggests only an outline of what we once learned, leaving many of the details to our imagination. As one researcher put it, "[Memory for] gist is enhanced by emotional arousal while memory for visual detail tends to be suppressed."[9] We unconsciously attempt to match the image component of an amygdala script (the gist memory) with our present situation. It is likely that this is because the hormones being released when a person undergoes a hurtful experience enhance the activity of the areas of the brain associated with amygdala-based memories and suppress those associated with declarative memories.[10]

In my brother's case, the outline for his image was of a rattlesnake poised to strike, but it was left to his imagination to automatically "fill in" such details as the size of the snake, the exact location of the snake, and whether or not the snake was actually coiled and facing him. The end result was that although he never saw the snake he "imaged" it as directly in front of him, average-sized, at the level of his neck, coiled, and ready to strike.

This is the nature of an amygdala script's image; it suggests an outline of a situation but not the content. It leaves the details up for grabs. For example, we might imagine a person in a social situation, a situation that unconsciously evokes previous social situations wherein that person felt humiliated. The unconsciously activated memory suggests that the present social situation represents a place where humiliation is in danger of occurring again, but the memory does not specify exactly how the humiliation is going to occur. It is left to the imagination of the holder of that memory to fill in the blanks: "These people won't like me and might reject me in some embarrassing way because I'm not intellectual enough" (or "I'm too intellectual"); "because I don't talk very much" (or "because I talk too much"); " because I don't have much of a sense of humor"; "they can see my nervousness"; "they are different than I am." The possibilities that lend themselves to fleshing out an activated amygdala image are almost endless. Once an amygdala script is evoked, the details of its image tend to follow.

So what is this nefarious part of our brain that solidifies the distortion of a script's image by "filling in the blanks"? What part of us would seemingly want to contribute to our undoing in this way—leaving us in a state of feeling and assumption that confuses us and alienates others? The likely culprit in this mystery turns out to be a favorite of the Western mind, the part of the brain that we count on for reality testing—namely, the neocortex. The neocortex is the part of the brain most capable of creating conscious assumptions about reality that could fill in the blanks of an amygdala script.

How could this be? How could it be that the conscious and more evaluative part of our brain rushes in to fill in the blanks, fleshing out a scripted image and making it appear to us that it fits into our present situation? From an evolutionary standpoint, this seeming betrayal by our neocortex actually makes a great deal of sense. Providing only a rough outline of a potentially dangerous situation has helped to assure our (and many other animals') survival. If the image of the rattlesnake that was evoked for my brother had arisen with the details already filled in, perhaps as an image of a snake that was to his left, he may have simply ducked to his right. If the snake had happened to be on his right, it could have been very unfortunate for my brother. To cite another example, if one of our ancestors saw a form that reminded her of a dangerous predator, one that had traumatized her before, and the image component of the amygdala script that arose was specific—for example, specifying what direction and by what means that animal was going to attack, an unconscious response based on those particulars could easily lead to disaster. On the other hand, a more generic image whereby the amygdala provides the gist of what is going on (you are in grave danger of being attacked by a vicious animal), and the neocortex fills in the details (it could come at you from somewhere ahead, no time to run, step behind that tree and freeze), then the scripted image might suffice quite well. As we have already seen, from the standpoint of evolution, it is no big deal should the assumption of imminent danger prove false.

It is worth emphasizing that when the neocortex is hijacked to fill in the details, no conscious evaluative function occurs about the origins of the emotional response. This has important implications: our mind automatically looks for and reinterprets available data *based on the assumption that the scripted "image" of what is happening is accurate.* The neocortex is not seeking information in the present situation that might contradict the image. It is seeking information that is *congruent* with the image. The result is that certain details that are congruent with the activated image are embraced and emphasized by our conscious mind, and other details in our present situation that do not fit the scripted image are minimized or discarded. There is an unfortunate lack of conscious consideration as to whether or not the scripted image of what is happening *is* actually happening.

A somewhat common (if a bit exaggerated) example of how this hijacking of the neocortex can occur demonstrates this. A person who is fearful of flying is sitting in the gate area waiting to board a flight. An amygdala-mediated fear suggests an image of danger associated with an airplane crashing. As he sits, he can see the plane. In fact, he has a bird's eye view of one of its wings. His eyes quickly scan the wing for cracks. Finding none, he nevertheless notes that the metal on the wing seems a bit old. In fact, the whole plane seems to have been in service for quite a while, and he wonders how exactly airlines do address metal fatigue. He now becomes aware of one of the attendants and notices that she appears unhappy . . . seems a bit upset, even—making him wonder if she may have some upsetting insiders' knowledge about his flight. A pilot appears, a rather young person, in fact, obviously not very experienced, hmm. . . . And with a jocular bearing that seems . . . well, a bit forced—like he also might be nervous about something. An older woman is sitting across from our passenger, she looks innocent enough until you consider that in these times of high security an innocent-looking older woman would be exactly the kind of person that a terrorist might want to recruit. Our passenger's mind quickly reviews the possibility that she has an incurable disease and has been offered a lot of money so that her heirs will never have to worry about money again. Very unlikely, he decides, but. . . .

Clearly, we could go on in this vein, until our prospective passenger decides that perhaps it would be better to fly another day and leaves. While our example may seem a bit exaggerated (especially to those readers who are not afraid of flying), it represents how a scripted imagery can hijack the neocortex into looking for and interpreting information in a fashion that supports the image component of a script. Most of us carry amygdala scripts that, at least to some degree, affect what we worry about, how we perceive the world, and how we behave: a woman mistakenly gives the cashier ten dollars too much and she blushes, is momentarily discombobulated, as if that type of mistake suggests that she is generally incompetent; an employee is absent-mindedly brusque with his boss, finds himself fearing various imagined consequences, eventually calls his boss to apologize, and discovers that his boss had not noticed anything; a man's wife is distant and distracted, and he immediately assumes that he has somehow upset her; a cop stops a man for a minor traffic infraction and the man becomes so anxious that his hands are shaking as if some terrible punishment is about to happen; a woman is mildly rebuked at work, and for the next two days she second-guesses what she is doing and doubts her ability to perform her job at all; a young person dies, and one person who knew the deceased continues to be haunted by guilty feelings, thinking that she hadn't been supportive enough of the deceased, even though her friends point out that she did a great deal to help this person; another person experiencing the same death continues to be bothered because he didn't seem to be as affected by the death as others were and worries that he might be cold and uncaring, while yet another person is left feeling worried and frightened by the death because it seems to verify an old image of the world being a dangerous place. Activated amygdala scripts are common. Although annoying or even embarrassing, they usually do not cause much of a problem. Most of us, however, have a script or two sufficiently intense and/or pervasive that they cause us problems.

The affliction of having our reality distorted by a script is compounded by an additional tendency of amygdala scripts—a tendency to prescribe "defensive" behaviors. An amygdala script often activates defensive behavioral responses that may or may not be in keeping with the immediate situation wherein the script was activated. Once activated, it is left to the neocortex to try to "rationalize" or make sense out of defensive behaviors. For example, people who feel shy and anxious in social situations (via an amygdala script that gets activated by such situations) may tend to blame others: "They're not very friendly and anyway they are really very superficial," or to blame themselves: "What is wrong with me? I'm such a social klutz."[11] Such rationales come easily to the hijacked neocortex, which after all, is charged with explaining whatever happens in "real" terms. In the throes of an amygdala script, we may be so caught up in blaming others for our feelings or so caught up in feeling bad about ourselves that it never crosses our mind that we are under the spell of an image that is old, habitual, and not relevant to our present circumstances.

The psychotherapeutic world has long been aware of the particular distortions that are associated with scripts. Psychotherapy began with

Freud—a physician—and it has since followed the medical practice of assigning different diagnoses to different categories of reality distortion that we can now see are scripted imagery.[12] For example, when people are depressed they tend to see themselves in a critical light or as helpless, and they may have images of the world as a punitive place wherein they are destined to be mistreated. Such scripted imagery is derived from historical experiences.[13] Once such imagery is activated, the brain looks for data to support it. The conscious mind locks onto aspects of reality that can be interpreted as supportive of this pessimistic view. The proverbial glass is perceived as half empty.

In the case of anxiety, imagery borrowed from earlier fear-inducing situations may suggest that the world is ripe with threatening pitfalls. A person who is carrying anxiety-producing scripts may (if the script is activated) interpret their present situation in a way that fits the assumption of danger. Psychotherapists have noted that these anxiety-related assumptions will tend to center around one, or sometimes two, of three common anxiety-reinforcing images: I am in physical danger (for example, I may have cancer or I may be in danger of dying of a heart-attack); I am in danger of losing control of my mind—going crazy; and last, but not least, I am in danger of humiliating myself and ending up a lonely misfit. In psychotherapy, these misinterpretations of our present circumstances have long been known to be the result of previous experiences—insight therapists refer to such distortions as historically based projections, and cognitive psychotherapists refer to them as cognitive distortions—distortions that were once learned and have now become habitual.[14] Our modern knowledge of brain physiology casts additional light on these distortions, allowing us to understand them as an image or gist that has been derived from previously painful situations.

We can now clearly see that scripts carry an image, and that it is inherent in the nature of scripts that when an image is activated, it willy-nilly imposes itself upon the situation we are in and does so with minimal review from the evaluative parts of our brain. More problematic yet, that image is unconsciously bolstered by the neocortical region of our brain, as it selectively focuses upon seemingly supportive evidence and creates distorted interpretations of real events. It usually does not occur to us that a particular, old, and influential imagery is imposing itself upon our present situation. However, by practicing mindfulness of body, we have seen that catching amygdala scripts in the act is a skill that is very attainable.

Unfortunately, practicing mindfulness and subsequently learning to be aware of when a script is activated often still leaves you at the behest of that script. For example, you may find yourself overreacting to something and apply mindfulness saying, "Yep, there's that old script again"—only to discover, to your chagrin, that knowing this does not take the power of the script away. You can be aware that a script is being activated and at the same time be dismayed to realize that your body and your brain are not nearly as impressed as you are by your newfound awareness. Your body may continue to produce that familiar sick feeling in your stomach or, in another instance, may give rise to a persistent angry feeling, or leave you stuck with a dark, heavy feeling in your chest, even while you are reassuring yourself that your reaction is accounted for by an old script. Try as you might, you may still have a hard time convincing yourself that the imagery of the old script does not fit.

There are two tools that, especially in combination, are very effective in allaying the persistence of a scripted image. One of these tools is *insight* and the other is *exposure*. Exposure deintensifies the emotional component. It is a one-two punch. The neocortex is called upon in a particular way to reality test (provide historical insight), permitting more judgment and reasoning to occur. At the same time, the reduced emotional intensity that results from the exposure exercise allows those areas of the prefrontal cortex that were formerly suppressed by a hormonal shower to become more free to operate normally.

Insight as a Therapeutic Tool

Beginning with Freud, there has been a long-standing emphasis in psychotherapy upon historical insight: the uncovering of the historical origins of the particular problem that harms us. As we noted in chapter 2, there is a substantial amount of research which suggests that indeed this kind of insight is helpful. When we apply insight to amygdala scripts, the insight tool involves discovering the early historical situations wherein an amygdala script was learned. This insight in return reinforces the mindfulness/awareness that indeed a script is being activated. Both the mindfulness that directly acknowledges the feeling component as it arises in the present moment and the insight that informs us that the image component is something that was learned in the past in turn pave the way for the effectiveness of the cognitive-change step (described in the next chapter), which addresses the belief component.

In the last chapter, we noted that recent discoveries in Western neuropsychology suggest the importance of practicing moderating scripts while they are activated. We added that this has greatly enhanced the effectiveness of psychotherapy. The most effective utilization of insight adheres to this principal of working with activated mental processes. Insight is most effective when it is used in the context of an amygdala script while that script is activated. Effective insight into the origin of a problem does not come about primarily because a person has had an insightful "Ah, ha!" experience. Effective insight is made effective because it culminates in repeatedly applying that insight to an activated script. In this way insight promotes mastery of a script.

Discovering Seed Images

The insight practice that we use in the second step requires that we identify the historical "seed images" that led to the creation of a script. We use a modified version of a technique that was created by Francine Shapiro to discover the historical basis for a script.[15] It is very effective and at the same time easy to learn.[16]

The exercise begins with step 1, described in chapter 4: evoking an amygdala script and tuning into its feeling component. (If you have not practiced doing this you may need to review the last part of chapter 4.) For many people, the exercise described below flows very easily. However, it is also possible that you will have to practice the exercise a few times before you get the hang of it.

We begin by remembering some recent situation that has evoked a script—a situation where your feelings did not seem to fit with what was going on. Perhaps you felt awkward, anxious, angry, upset, depressed, sad, confused, or some other feeling that was problematic and, in retrospect at least, seemed inappropriate or disproportionate to the situation. If you need to, use the decision tree presented in chapter 4 to ascertain whether or not what was evoked was a script.

Make your memory of that situation as detailed and as vivid as possible until you notice that emotions are being activated. Notice what parts of your body you associate with the emotions, paying special attention to your torso. You do not have to be perfect at this. Just do your best at trying to notice the emotions that are being activated. Think of these feelings as occurring in your body. If it turns out that you can only think of emotions as occurring in your head, don't worry about it. Focus on them there.

Stay focused upon the emotions that are associated with the script that was activated in the recent situation that you are imagining. If your mind wanders, come back to an image of the evocative situation and then focus on the physical sensations activated by the emotions. Spend a minute or two simply maintaining that focus.

As you actively maintain your focus on the emotions, your mind will tend to passively drift from image to image of times when you have felt the same feeling in the past. Let your mind wander in this way, as if you are watching a slide show of events where you felt this same or a very similar emotion. The images do not need to be accurate historical images. If they are vague and arise without much detail, that's fine. Do not force it or actively seek out these memories. Trust that your mind has a natural tendency to go from image to image of previous times when you have had similar feelings in the past. Simply suggest to yourself that you have had these feelings in the past and watch where your mind takes you.

Do not evaluate what comes up. For example, do not try to figure out if it really happened the way that you are remembering it. We are not interested so much in historical accuracy as we are in images that depict what you once felt like. Also, avoid getting caught up in questioning why a particular image comes up and what its relationship is with the present situation.

Try to not attach special importance to what arises. Stay with the feelings and watch the imagery with a mild and somewhat detached attitude. If you start getting into story lines about what happened or begin to formulate essays in your mind about the significance of what you are remembering, drop them and come back to the feelings in your body.

If you get distracted, come back to any one of the images that re-evokes the script and focus on the feelings associated with it.

Let your mind drift from occasion to occasion when you have felt this emotion, going further and further back into your history. Do not work too hard at this; actively keep the focus on the feelings and passively notice the images of past times when you have felt this way. Do not guess at what will come up next. Simply watch what arises.

Eventually you will find that particular images associated with a particular time will remain and other earlier images won't seem to be arising. After some time you may decide that you have reached a "seed image," an image that depicts what was going on during at least one early juncture where you were learning the particular script on which you are focusing.

Sometimes people report that a seed image is very vague. That's fine. To flesh the image out a bit, ask yourself approximately at what age it feels like you were when the vague image was occurring. Then ask yourself where it seems like you are: in your house, in the yard, at school, with friends, and so on. Then ask yourself if anyone else is present. Usually this sort of questioning helps define an image. It should be emphasized that we are not concerned about whether or not the details of an event actually happened the way that the imagery is depicting it. We are only interested in imagery, a picture that depicts the environment you were in when you were learning a script.

It is not unusual for people to settle on a seed image only to doubt it later. If that is the case, do the exercise a few times and see if the results are the same. If so, you can begin to trust that what you are coming up with as a seed image is important and relevant to how you learned a particular amygdala script. If you are not sure about a seed image even though it keeps coming up in the above exercise, go ahead and work with it in the ways we describe below. By doing that, one of two things will happen: you will either begin to see more clearly how it is relevant, or your mind will naturally find itself drawn to other images with more emotional strength, and you can focus on those other images instead.

Janet's story is a typical example of how this exercise can work. Janet came into therapy complaining of depression and anxiety. She focused much of her first two sessions upon her marriage. She was in a marriage that was not working. For years, she and her husband had maintained a relationship of polite cooperation. He did not appear happy, and was politely disinterested in her at best—and she was clearly miserable. Janet's work took her on the road a lot and she had had a couple of brief affairs, but they were very unsatisfying. She had tried individual therapy before and also couple therapy, but it had not seemed to help very much. She continued to make some attempts to discuss with her husband how the marriage was not working for her, but his response was calmly dismissive—it was working well enough for him. This would leave Janet feeling guilty, anxious, and self-critical, effectively ending their conversation.

Janet's previous psychotherapist had wanted to focus on her dependency needs, which he assumed underlay her difficulty—either her difficulty in separating from her husband or in risking his rejection by more effectively standing up for herself. In her third session with one of our therapists, Janet practiced the exercise described above, and the seed image that came up for her did not seem to have much to do with dependency. Hers was an image of being eight years old at the dinner table with her parents and her older stepbrother. In this image, it seemed that whatever she said was only politely acknowledged, while whatever her step-brother said was listened to with interest. The image included Janet letting her family know in various ways that she was unhappy and that she wanted more attention and validation from her family, but her family's response to her was more criticism and more discounting. (Janet actually had a series of images along the same lines, but the one described above seemed to epitomize the others.) In her present life the seed image that was getting evoked for Janet was one where her feelings and needs were not going to be attended to. Her husband's comments that the relationship felt fine to him was a signal to Janet that her feelings were off, and not important. At these and other times, it seemed that what she experienced or thought was not considered by others to be worthy of serious consideration. Furthermore, when in the grip of this script, it seemed to Janet as if it were true—she assumed that her frustration with her marriage was wrong. She had learned to habitually marginalize her feelings and her desires.

Janet's discovery of the historical origins of her difficulties is typical of the clients with whom we work. Her present therapist had trusted her to find the historical imagery that was relevant to her. Her discovery was a bit surprising to her. She had always remembered her early history as being idyllic, and in most ways she was right: it had been a good childhood. In fact, a laborious analytic process of examining her history might not have yielded much. Her insights occurred through the natural process described above where she allowed her emotions to lead the way to her insight.

It has been shown that the amygdala is especially active in children, so often the exercise described above leads to seed images that are associated with childhood. However, seed images can be traced back to a variety of ages. It is not unusual for people who are using this technique to find a number of different images associated with different times in their personal histories that are sufficiently emotional that they can be seen to serve as seed images for a particular script. This phenomenon has been reported by Shapiro.[17] Humans, when confronted with emotionally laden experiences (for example, a video of a tragic accident) take on new amygdala-mediated learning and retain a high degree of recall for the gist of those experiences for a long period of time.[18] An already activated amygdala script makes us especially susceptible to additional traumatic input. Thus, when following the emotion component of a script back to seed imagery, it is not unusual for people to come upon a few "stopping off" points: images of junctures associated with especially strong feelings that represent times when an already activated script became further intensified. People generally find it easy to recognize these images and to identify them as depicting times when their script was further fueled by external events.

Earl, for example, learned a script when his parents divorced. He came into therapy because he was having difficulty addressing issues with his girlfriend. When they tried to talk about conflicts that they were having, he would withdraw into a confused, anxious, and sullen silence. The seed image that arose for him was when he was eleven years old and he was leaving his mother's house to spend a week with his father. Immediately after leaving the house, he realized that he had forgotten something, and upon returning to the house he found his mother sobbing uncontrollably. The intensity of her pain was preserved in the script he brought into therapy. By tuning into the feelings that were part of a script that was hamstringing him in his present relationship, he was brought back to that moment. After he worked through that first seed image, however, he then found himself focusing upon two other images, also quite emotional and associated with the same script. In one, he had just told his junior high girlfriend that he was breaking up with her, and when she went running from him sobbing, he felt devastated by the same feelings of guilt and confusion that he had experienced with his mother. In college, he had had another relationship where a woman he was dating dropped him after a few months, saying that she was very frustrated with his lack of interest in her. Again he felt very strong feelings of guilt and anxiety. A history of repeated activation and reinforcement of scripts such as Earl experienced is common. Happily, once we have worked on the first seed image, the others become easier.

In choosing which seed image to focus on first for doing the practices that follow, two considerations are important. First, it is best to begin with earlier images and then work chronologically forward to later images; and second, it is best to start with the images that have the most emotional intensity. While these may be competing considerations, after a little bit of experimenting with a seed image it quickly becomes obvious which of these two variables is more important in defining your first focus. For example, if you have decided to focus on an earlier seed image and find that your emotions keep drawing you to a later image, try working with the later image first.

Most of the seed images that a person identifies (the early painful events that resulted in his or her script) are easily associated with actual historical occurrences. However, it is well documented that those memories which are processed, with little amygdala involvement, through the hippocampal part of the brain ("declarative memories") can include a variety of detailed information, whereas emotion memories that are processed with notable amygdala involvement (that is, amygdala scripts) lose historical detail.[19] If a seed image portrays a situation that you have not previously remembered, it is especially noteworthy that seed images should not be thought of as being historically accurate. What you are remembering is an image that depicts your inner state when a script was being established—in research parlance, you are calling up the memory of the "gist" of what happened, or an "implict" as compared to a "declarative" memory. A declarative memory more accurately depicts historical detail. So, while it is probably an accurate depiction of what you were feeling at the time, the particular historical events associated with a seed memory may not be so accurate. You do not have to assume that they are not accurate; they may well be. Just do not make too much of it as historical truth. This is why we usually refer to seed *images,* not seed *incidents.*

This caution regarding historical accuracy is especially important if the remembered image comes to you as a shock and depicts important events for which you have not previously had a memory. These types of memories are called "recovered memories," and they have been shown to be unreliable from the standpoint of historical validity. Some recovered memories have been shown to be accurate, others have been demonstrated to be false, and no way of distinguishing between the two has been discovered.[20] So, unless the seed image that your mind has settled upon is a familiar part of your history or fits with the memories of others, do not make too much of it as an historical fact—should you be writing your autobiography, or talking to relatives, for example.

My brother's experience with the rattlesnake portrays another interesting point. Although my brother has encountered rattlesnakes a number of times, he has never been harmed by one. In other words, the activating image of a snake about to strike and the fear that my brother felt were not the result of an amygdala script that had been instituted by his direct historical experience. Instead, his script regarding rattlesnakes was undoubtedly the result of growing up in desert country where the repeated childhood admonitions from our parents, along with various tales we had heard and pictures that we had seen, had instilled in us a fear that had become conditioned to the sound of rattling.

This is analogous to what we sometimes discover about seed memories. Sometimes seed images do not include something painful that acts directly on the person who carries the script. We have already seen this with Earl, who was traumatized by his mother's pain. In fact, Earl's experience, in which a child takes on a script as the result of empathy with the pain of a parent, is a common seed image. What was traumatizing was not a person's own experience of being hurt, it was their experience of their parent's pain. Another not uncommon instance of script-making that comes via the feelings of another is when a sibling observes another sibling or a pet being traumatized.

Clients who have anxiety reactions frequently come up with seed experiences that are as much the result of the reactions of others (often their parents) as they are the result of their own feelings. There are important variations of this that are common with anxiety problems.

Jed's experience typifies one way in which this can happen. Jed was born sensitive, what Carl Jung has called a feeling-oriented person. The seed image that came up for Jed was an image of going through a long, dark tunnel in a car with his parents. When he became afraid, his parents, whom he remembers as already being upset about something else, became very irritated and critical of him.[21] As a sensitive child, it is likely that Jed was born with something of a hair trigger, such that his body was more prone than most to react to situations by creating a surge of adrenaline. The combination of the suddenly darkened interior of the car and the tension that existed between his parents triggered a rush of adrenaline into his body. His parents' subsequent critical reaction to his fear made it impossible for him to mediate consciously between what was occurring in his body and what the realistic external situation was. Quite the contrary, their angry, critical reaction to his fear added to his fear and also taught him to feel as if there was something wrong with his fear. In light of his parents' reaction, young Jed felt fear times two. He not only felt fearful of the tunnel but he also felt as if there was something wrong with him feeling fearful—he now felt self-critical and fearful of his fear. As he later noted, these feelings became inseparable from the anger and tension that existed between his parents. Similar reactions by his parents to Jed's fearful feelings occurred repeatedly. He learned to associate his fear with something that was bad and unacceptable. As is typical of people who have difficulty with anxiety, he learned to fear his fear and to be critical of himself whenever his body dumped some adrenaline, something that his body was prone to do.

Reducing the Intensity of the Emotional Component of a Script

We now turn to the tools used to reduce the intensity of the emotional component of an amygdala script.[22] As we have already noted, reducing the intensity of an emotional component is synonymous with reducing the particular hormonal deluge that is the basis for that emotional component. In chapter 1, we mentioned that intense hormonal output affects the prefrontal cortex by suppressing parts of it, with the result that we do not consciously and rationally relate to our environment. By reducing the hormonal output that inhibits the prefrontal cortex (that is, by reducing the intensity of the feeling component of an amygdala script), we free up the consciousness-producing, reality-testing part of our brain, making it more open to new learning, more open to creating a different relationship with an activated amygdala script. More specifically, reducing the intensity of the emotional component of a script makes step 2 and especially step 3 (described in the next chapter) much easier. Once the strength of the emotional component of a script is lessened, it is usually not necessary to continue to practice the exposure exercise described below. (In some cases an occasional repeat of the exposure practice is necessary.)

_____The Discovery of Exposure as a Way to Modulate Emotion

Exposure is a psychotherapy tool that was mentioned in the first chapter. Exposure therapy had its beginnings in the 1920s, when a famous psychologist performed a very unethical experiment. John Watson was the psychologist. The experiment began when he gave a small boy— referred to as "Little Albert"—a pet rat. After awhile, each time that Little Albert touched the rat a loud and very frightening gong sounded. The boy quickly learned to fear the rat. Watson was able to demonstrate that the boy had become fearful of all different types of fur or fur-like things. Even though Albert had originally showed no fear of fur, he was now "fur phobic." Watson was planning to demonstrate that if the boy was repeatedly forced to be near furry things and there were no negative consequences for this (no loud gong sounds), the boy would get over his "fur phobia." The boy moved, and so the experiment ended. Nevertheless, Watson published his findings. Mary Jones, another psychologist at the time, took note and demonstrated with a different fur-phobic child (who we are happy to note had acquired his fur phobia naturally), that indeed Watson was correct—by repeatedly introducing this second child to furry things, with no associated negative consequences, then the phobia disappeared.[23]

Over the next couple of decades, the process for reducing negative emotional reactions continued to be studied: it was concluded that if a person remained attuned to a painful emotional experience, after awhile the intensity of the emotion would become markedly reduced.[24] However, these early attempts to make what we now know as exposure therapy a mainstay in the psychotherapeutic world met with only limited success. Although exposure therapy was still able to generate some interest, especially in regard to its effectiveness with fear-related problems, it was mostly relegated to a back-seat status in the world of psychotherapy.[25] This began to change in the 1980s, when Edna Foa refined the approach, added a cognitive component to it, and demonstrated that this combination was very effective in the treatment of post-traumatic stress disorders (PTSD).[26] Largely as a result of these refinements, exposure therapy has regained a great deal of status in today's psychotherapy world, and has now been shown to be an effective treatment component for a variety of diagnostic categories.

There are numerous protocols for applying exposure techniques to reduce emotional intensity. They all begin with arousing the targeted emotions. For example, Edna Foa in her work with PTSD clients has them tell a detailed story of the situation in which they were traumatized—for example a sexual assault or an auto accident—whereby they evoke the image component of a script. Her clients are instructed to tell the story again and again in a detailed fashion.[27] One of the more important refinements that Foa used to improve the effectiveness of exposure therapy was to include a prolonged activation of physical sensations associated with emotional states.[28] Shapiro has included a focus upon physical sensations in her utilization of an exposure process that is one of the mainstays of Eye Movement Desensitization and Reprocessing (EMDR).[29] EMDR has been finding its way into a variety of therapeutic arenas, and the way in which we utilize exposure therapy here is similar to Shapiro's approach.[30]

The best way to learn to use exposure to reduce the intensity of an amygdala script is to experience how it works. Before we do this, however, a few cautions are in order. The vast majority of people whom we have encouraged to practice the following exercises have no problem with it. Nevertheless, we want to encourage you to use common sense. Exposure treatments obviously have to do with evoking feelings, and if you have a serious heart condition or other serious cardiovascular problem, such that your doctor has suggested that you should avoid "getting excited," then you should either consider not doing this exercise, seek out a consultation with a professional who is well versed in exposure therapy, or check in with your doctor first. If you have a history of psychosis or extreme manic states, then it would be best to not do these exercises or to do them only under the guidance of a licensed psychotherapist, and then only after careful consideration. Rarely, people who do these exercises experience a strong residual reaction to them; perhaps they feel dissociated for awhile, or they feel very emotional and it takes a few hours to shake it off. If this is the case, then it would probably be wise to discontinue working with the exposure tool for now. Below you will find a discussion of how to use a relaxation exercise as an alternative method for reducing emotional intensity. After practicing the relaxation exercise for a few weeks and after working with the three steps for a while, you may want to again attempt the exposure piece. If you continue to feel overwhelmed or dissociated and wish to continue working with a script, you should seek out the help of a psychotherapist. Hundreds of people have practiced these exercises on their own and very few have had any serious problems with them, and the problems they did have were almost always short-lived. But do use your common sense. With what we are presenting here, as with any model of psychological growth, if you feel stuck or feel worse while practicing it, it is advisable to seek help from a licensed professional psychotherapist.

The Exposure Exercise

When practicing the exposure exercise make sure that you have plenty of time. An hour and a half, for example—it probably will not take that long, but better too much time than too little. Also, make sure that you will not be interrupted. It is okay to have a trusted, supportive friend or spouse with you or in the house with you—in fact you may prefer this—but let them know what you are doing and that you prefer that your process not be interrupted.

———

Begin by using the procedure we have described above to evoke a script and to identify a "seed image"—a historical situation that exemplifies how you learned the script. Once you have identified a "seed image," begin the exercise by recalling a recent situation that evokes the script you want to focus upon. As the old emotions associated with the script begin to arise, notice how they feel in your body and say, "There's that feeling." Then take your mind directly to the seed image. Try to do this while keeping your eyes open. The reason for having your eyes open is that this reminds you that the feelings you are working with represent what happened in the past and you are in a different place and a different time now. However, if having your eyes open makes it difficult to concentrate, experiment with closing your eyes, making sure that you can do this without losing track of where you are.

As you call upon the seed image, try making that image vivid enough that you can notice the emotions associated with it. Focus on those emotions as sensations that exist in your body. If you can associate those sensations with particular areas in your body, especially your torso, focus your attention there.

When your mind wanders, bring yourself back to the image and focus upon the feelings in your body. Every few minutes, remind yourself that what you are experiencing is an old learned emotion memory that does not fit your present circumstances. This is a very important part of the exercise—the constant conscious reminders that the feeling is simply a memory of something that happened a long time ago.

After a few minutes, ask yourself about the intensity of the feeling. Ask yourself how you would rate the intensity of the feeling on an intensity scale of one to ten: "ten" being over the top and seemingly unbearable, and "one" being "There's a feeling but it is no big deal at all." Rating your feelings in this way establishes a baseline marker of how intense the feeling is.

Continue to maintain a focus upon the feeling, watching it. Keep doing this, calling your mind back to the seed situation, and focusing upon the feelings and where in your body you associate that feeling. Keep reminding yourself, even saying to yourself, "This is an old historical feeling that I learned in the past. It has no real relevance to the present." Eventually, you will notice that the intensity of the emotions drops off. Typically, the intensity bounces around a bit, but eventually it will even out and gradually come down. When it gets down to a "three" or better yet a "two" and stays there for five minutes or so, take a break. Get a drink of water maybe. Or walk around for a few minutes. Then do the exercise again to make sure. After the first time you have reached a "two" or "three" it will probably take significantly less time to get back down to those levels of intensity. In fact, often people find that when they return to the exercise, they begin at a two, three, or four.

––––––––

Repeat this exercise a few times over the next few days. After the first or second time that you do it, you will probably find that it takes much less time, perhaps only a half hour, and after that maybe just a few minutes. When this is the case, you have successfully reduced the intensity of the feeling.

_____Relaxation Exercise to Reduce Emotional Intensity

In appendix A, you will find a detailed description for practicing a very effective relaxation exercise. This is another way to reduce the intensity of the emotional component of a script. From the standpoint of amygdala scripts, a daily practice of relaxation is especially useful in those cases mentioned above in which the exposure exercise either does not work or is not advisable.

Practicing an effective relaxation methodology for fifteen minutes, three times a day, has two benefits. First, a daily relaxation practice reduces your baseline level of arousal or tension. Different people operate at different levels of arousal as they go about their daily lives. The higher the arousal rate, the closer one is to reaching the threshold for stress or anxiety. By instituting a relaxation practice into your life, the base level of arousal lowers, and so this threshold is less likely to be reached. Also, a relaxed body is less likely to be easily triggered by a script, and when a script *is* triggered, it is easier to be mindful of it and it is easier to be skillful with it. In short, when a script is activated and you have been practicing an effective relaxation exercise, the script will usually be less intense and it will be more amenable to being mastered.

The second benefit derived from a regular relaxation practice is that you become very good at it. Trying to attempt to relax yourself in the face of an amygdala script is made much easier if you have been practicing relaxing yourself throughout your day and have been doing this regularly for many days. Trying to apply a relaxation tool to an activated amygdala tool without such practice is less likely to succeed. The daily practice of a relaxation exercise is especially helpful if the feeling component of a script includes a significant amount of anxiety.

_____Step 2 in Mastering Amygdala Scripts

Once we have identified a seed image as our target and, if necessary, have reduced the intensity of the emotions associated with that image, we are ready to practice step 2 of the Three-Step Practice. Step 2 is now very simple. It entails calling up an early seed image and saying to your self, "That's where I learned this feeling." It's as simple as that.

So at this point we have step 1—evoking a script, tuning into the feeling component of that script and then saying to yourself, "There's that feeling." Step 2 follows when you flash back to a seed image of the script and say to yourself, "That's where I learned it." By calling up a seed image that represents where you learned that feeling, along with saying, "That's where I learned it," you have extended your conscious awareness of the script to include recognizing it as a particular emotion memory that you learned a long time ago. The emotional imagery of a script has changed from something that seems true in the present to become a historical memory with relevance only in the past.

Like step 1, there is also a mindfulness aspect to step 2. Clients have described the experience of steps 1 and 2 as "creating space around a script." These steps provide a vantage point from which it is possible to step back and look at a script—see it in its true form as something that simply arises in your mind. Unacknowledged and left to do their own bidding, scripts do what they were designed to do—raise an alarm and narrowly focus all of our attention on an image of potential harm. In doing so, scripts lead you to ignore where you are and what is really most real: perhaps, the birds that might be heard singing outside; the play of light, color, and texture that surrounds you; the feeling of the cloth of the chair on which you may be sitting; and many other things which are truly more real than the images that are arising in your mind. As people practice step 1 and step 2, separating themselves but not suppressing the feelings and imagery of their script, they free themselves to experience more space and perspective, and greater attunement to their environment and to the people around them.

We are not quite through yet, however. We have a very important third step to add to the amygdala mastery practice. Nevertheless, it is worthwhile to stop and give yourself some credit for what has been accomplished already. You have learned that in order to work most effectively with your mind, it is important to engage it as a process and then to work with the process as it is happening. It is no small accomplishment to notice those things that you have unconsciously been avoiding for years and to practice stepping into embracing, rather than ignoring or fighting with problematic feelings and images. Most important, by experimenting with the first two steps, you have not only theorized about this but you have actually begun to experience identifying a script, calling it up and exploring it, even reducing the intensity of it.

Perhaps you have begun to notice that working with even the most problematic aspects of your mind entails just that—working *with* your mind rather than dividing it up into adversarial parts. Way too often we have been taught to fear our minds—to approach the less conscious parts of ourselves with trepidation. This has been reinforced by a miscasting of the unconscious as a dark and dangerous place. Similarly, we have been taught to believe that our psychological problems are the result of something that is broken in us that should be rejected. One of the most important things that you can learn by experimenting with amygdala scripts is that you can afford to embrace and work fearlessly with your mind. Rather than threatening you, your mind can become an ally that frees and supports you.

_____Step 3

CHANGING THE
BELIEF COMPONENT

When Martha's script is activated she practices the third step by imagining being in a seed image with her twelve-year-old former self—a girl who had just been told that she was an "ice queen," selfish, mean, and would go to hell. In this fantasy, Martha imagines that she puts her arm around this little girl, assuring her that adult Martha completely understands why she feels so bad. She lets little Martha know that even though she may have done something wrong, she is nevertheless a fundamentally good-hearted person and that the awfulness she has just experienced occurred because her parents handled the situation very poorly. Martha then imagines her twelve-year-old self taking this in and feeling much better.

In the seed image that Mike is working on, he is an eight year old who has just been hit so hard by his father that the handle of a wooden paddle has broken. Mike projects himself into that situation and tells his father that no matter what little Mike had done, it did not warrant being treated so brutally. He then imagines empathizing with how bad the eight year old feels and reassures young Mike that in the future he would prove to be a responsible and loveable person. Mike then imagines that those words allowed little Mike to feel like someone finally understands him and accepts him, imbuing the little boy with a sense of warmth and well-being.

Stephen's seed image is of himself, ten years old in the kitchen with his mother, who is crying and devastated because his father has just moved out. Young Stephen is about to go visit his father and he feels responsible for his mother's pain. He believes that he should be able to help her, but finds himself hopelessly inadequate for the task. Adult Stephen imagines himself appearing in the seed image and letting his ten-year-old self know that he understands how helpless, frightened, and worried little Stephen feels. In this image, adult Stephen tells ten-year-old Stephen that, while it is great that he is so good-hearted, realistically he is not capable of nor expected to relieve his mother's pain—simply empathizing with her is enough. Adult Stephen then imagines ten-year-old Stephen feeling relieved and understood.

Geri imagines herself as the seven-year-old kid sister she once was, and her older sister has just ridiculed her in front of her friends. Geri practices the third step by inserting her adult self into this image, letting young Geri know that she understands, in fact is feeling in her own body, how disappointed, crushed, and all-around awful the seven-year-old feels. She then puts her arm around young Geri, letting her know that nothing is wrong with her—that she is in fact loveable.

____Discovering the Belief Component

The examples of Martha, Mike, Stephen, and Geri portray how seed images can be used to work with the belief component of an amygdala script. By evoking and inserting herself into her seed image, Martha highlighted for herself an unconscious belief that had often undermined her self-confidence: she is selfish, mean, and in danger of going to hell. Mike's image of being brutally hit by his father allowed him to further access the anger and fear that he felt at that time and to realize that while he was being hit, his mind was recording a belief that he must be such a bad person he deserved to be treated brutally. Stephen's image of his devastated mother's grief elicited strong feelings of sadness and helplessness in him, feelings that bolstered what seemed to be "proof positive" that he was useless to those whom he loved, especially when they were distressed or going through a hard time. The humiliation and confusion that Geri felt with her older sister instilled in her the belief that she was socially unacceptable—a real loser.

____Identifying the Belief Component

A straightforward way to uncover the belief component of a script begins with recalling a seed image (part of step 2) and then tuning in as much as possible to the emotions that you had in the seed-image situation. After this, if you ask yourself, "What seemed true about me at that time?" you will probably discover a general outline of the belief component of an amygdala script. Refining the belief component into language that is the most accurate fit may take a bit of experimentation. The questions: "What seemed true about myself while this (the seed image) was happening?" or "What was the message I was taking in about myself at the time of the seed image?" yield the first renditions of a belief component. As we continue to practice with a script, we may find ourselves modifying our answer to these questions. Soon we discover a belief that seems to be a good or a "good enough" fit. Neurologically, when you ask yourself these questions you are calling upon the neocortical, consciousness-producing parts of your brain. As you practice working with the belief component, you condition this additional pathway to the neocortex, further ameliorating the effects of your amygdala script.

Another benefit from identifying a belief component is that it provides an additional means for tuning into your script—the belief component serves as a handy label for a particular script. Steve came to refer to his script as "the caretaker who is afraid to let people hurt." Martha referred to her script as "my fear that no one will like me"; for Mike it was "the part of me that feels worthless"; and for Geri, "the part of me that thinks I'm a loser." A ready label for identifying a script makes it easier to be mindful when it is activated.

Once we have identified the dysfunctional belief component of a script, we feel compelled to change it. Cognitive therapy seems like an exceptionally good approach to this, since it is a therapy that is predicated upon identifying and changing dysfunctional beliefs. We suggested in chapter 1 that although it has not caught the public imagination (and indignation) in the way that Freud's theories once did, cognitive therapy has nevertheless had an impact on the practice of psychotherapy that is arguably as powerful and far-reaching as that which Freud prescribed a hundred years ago. A brief review of the origins and basis for cognitive therapy will help us understand how the practice of the third step works and why it has taken the form that it has.

____The Origins of Cognitive Therapy

In the 1970s and 1980s, a virtual tidal wave of research into cognitive therapy forever changed how psychotherapy is done in the West. The two central figures in this revolution—Albert Ellis and Aaron T. Beck—both had their beginnings in Freudian psychoanalysis, but their work, and especially the research that was spawned by their theories, resulted in a dramatic departure from Freudian theory.

This revolution in thinking began early, in the 1950s, when Albert Ellis suggested that the accuracy of our internal dialogue—how rationally or irrationally we talk to ourselves—represents a powerful leverage point for taking control of our mental health.

His idea caused little stir, probably because it seemed too simple. In fact, most therapists at that time assumed that such an idea was simplistic. Ellis, however, is one of the most indomitable and forceful personalities in the history of psychology. He persisted in refining and experimenting with "dysfunctional thought patterns" and eventually developed a means of doing therapy that centered around correcting a variety of irrational assumptions that commonly exist in people's habitual thought patterns. He originally referred to this form of therapy as "Rational Emotive Therapy" but RET later evolved into the Cognitive Therapy or Cognitive Behavioral Therapy schools that we have mentioned in previous chapters.

By the 1980s, due largely to the influence and modifications provided by Beck, cognitive therapy evolved into a well-formulated, broadly applied, and powerful form of therapy that was pervasively influencing the world of psychotherapy. Presently, cognitive therapy has been integrated into almost all of the actively researched theories and methods of psychotherapy that exist today.[1] The reason for the great esteem that cognitive therapy quickly garnered is simple—it was strongly supported by rigorous research.[2] Beginning in the late 1970s and up to the present, the most highly regarded psychotherapy research annals have been dominated by cognitive therapy studies that show its effectiveness. Other approaches to psychotherapy have also continued to prove themselves, but cognitive behavioral therapy has maintained its status, so that it now is the standard by which other therapies measure themselves. At present it is common to explore ways of improving earlier therapy approaches by integrating more cognitive therapy into them.

____The Nuts and Bolts of Cognitive Therapy

So what exactly is cognitive therapy? Cognitive therapy does not concern itself with complicated interactions between various psychological phenomena such as developmental stages, personal histories, antisocial impulses, and unconscious versus conscious aspects of the mind (to name but a few). Cognitive therapy focuses on human thought processes. The proven effectiveness of cognitive therapy clearly demonstrates something that previously had not been well appreciated—simply changing how people habitually think has huge consequences for how they feel and behave. Our minds tend to be filled with seemingly continuous subconscious chatter or dialogue. This "internal dialogue" constitutes the mental river that sculpts our assumptions about our self and the world. We then experience the world through the lens of those assumptions. A cognitive therapist teaches patients to identify components of their internal dialogue and then encourages them to experiment with discovering how powerfully those assumptions affect them. The next step is to create a more satisfying psychological world by teaching patients a variety of methods through which the assumptions can be changed.

To experience what we mean by "internal dialogue," try this experiment. Focus your attention on your breathing. Pay attention to every out-breath as it leaves your body. Imagine that once the out-breath has left your nostrils it dissolves into the air around you. Now decide that, for the next five minutes, you will simply maintain the focus of your attention upon the out-breath—for the next five minutes the only thing that you will do with your mind is imagine your out-breath. Try it, now.

Unless you are an experienced meditator, and assuming that you carefully observed what was going on in your mind, after a few seconds of trying to maintain your focus on your breath, you will have found that your mind became distracted from your breath. Perhaps you began thinking: "What is this silly exercise about?" or, "Why am I doing this?" or, "Am I doing this right?" or, "Don't I have something better to do now? Like maybe wash the dishes?" A bit of experimenting with the above exercise will demonstrate a long list of the many ways in which your mind manufactures thoughts and feelings—pretty much an uninterrupted and endless parade of mental phenomena.

The flow of mental activity that we notice in this exercise is what is meant by "internal dialogue." Cognitive therapy focuses exclusively on the thought component of this flow, relegating the imagery and feelings to secondary importance. Cognitive therapy seems to have demonstrated that by changing the content of some of those thoughts, we can positively influence our moods, our perceptions of others and, most important, our perceptions of ourselves. However, neurobiology has uncovered some clues that suggest that some of the things that powerfully contribute to the effectiveness of cognitive therapy have nothing to do with modified beliefs.

____A Different View of Cognitive Therapy

We have already noted that as we repeat the Three-Step Practice, from a neurobiological standpoint we are enhancing ("potentiating") pathways between the neuromechanisms of a particular script and the consciousness-producing areas of the neocortex through a conditioning process. We have also noted that in order to tame a script effectively these pathways must be engaged while a script is active. This requires repeatedly evoking a script and then changing how we relate to it. Without this conditioning, a script will tend to continue on its not-so-merry way despite the protestations of consciousness.

This brings us to an intriguing possibility that we touched on briefly in chapter 4. Perhaps the effectiveness of cognitive therapy does not result primarily from new beliefs being instituted to replace old dysfunctional beliefs. It could be that the benefits are largely the result of neocortical regions of the brain being repeatedly activated when we attempt to change the content of an old belief. Analogously, when we embark on an exercise regime, say climbing up a hill three days a week for thirty minutes, the benefit that we derive from that exercise results from the climbing, not from whether we reach the top of the hill. Could cognitive therapy be similar? The process of cognitive therapy necessitates repeatedly observing one's thoughts and identifying the target belief as it arises. This means repeatedly enhancing different pathways from the amygdala/limbic system areas of the brain to the consciousness-producing regions of the neocortex. Conscious awareness becomes conditioned to arise when certain thought patterns are present. Once such conditioning has been established, the rational and evaluative regions of the brain are engaged to moderate an active script. Contrary to what cognitive therapists have assumed, the crucial conditioning process may not depend on what the new belief constitutes or even whether or not there is a new belief. Instead, it seems plausible that much of the power of cognitive therapy is the result of the process it entails, not the displacement of dysfunctional thoughts with more functional self-statements.

If this hypothesis is true, it would have implications for further streamlining the three steps (and important implications for cognitive therapists, in general). Hypothetically, when a person recognizes the belief component of an activated script, it might be as useful for him to think of, say for instance, "ice cream" to remind himself of a changed belief. Or, from a slightly different angle, it might suffice simply to label an old belief, consciously acknowledge it as something that is arising in one's mind, and not even worry about changing it. Such a practice of conscious acknowledgment can be readily seen to be a mindfulness practice (akin to the Buddhist practice of "mindfulness of mind" where the object of mindfulness is a thought, not, as in chapter 4, an emotion). Recently, there has been strong and growing evidence that supports just such a view—that mindfulness is a major, probably even a primary factor, in how cognitive therapy really works. For example, Teasdale and colleagues concluded in their research that cognitive therapy "may reduce relapse [of depression] by changing relationships to negative thoughts rather than by changing belief in thought content."[3] They acknowledge that the method that they identified to account for this change was the result of a mindfulness process inherent in cognitive therapy: a process that creates a "cognitive set in which negative thoughts and feelings are seen as passing mental events rather than as aspects of self."[4] We can add mindfulness of thoughts as a very important dimension that contributes to the effectiveness of cognitive therapy. There is yet another, a third ingredient, that is a major contributor to the success of cognitive therapy (and to all other therapies, for that matter). This well-known contributor to personal change adds a third dimension to the cognitive-change strategy that we have included in the third step.

How Empathy and Warmth Contribute to Personal Change

When, after twenty years of doing psychotherapy, I decided to more fully engage the tools of cognitive therapy, I began to attend seminars of those who were foremost in the field. Two of these people especially impressed me—Geoffrey Young and Christine Padesky. In addition to their depth and breadth of knowledge, I was struck by how strongly they encouraged their audience of therapists to adopt a warm and accepting attitude toward their patients. This message was underlined by the degree to which they themselves portrayed a caring, empathetic concern in taped sessions with their patients. Even though caring and acceptance are not intrinsic to cognitive therapy theory, these masters of the technique clearly considered those qualities to be centrally important.

Intuitively the value of warmth and acceptance probably seems obvious. Ask a good parent or grandparent, "What's the most important thing that a parent can offer to enhance the psychological well-being of their child?" and most will, in short order, find themselves talking about the child feeling loved and accepted.[5] So it is not surprising that genuine empathy and warmth (or "unconditional positive regard") have been shown to play a central role for positive outcomes in psychotherapy.[6] This is as true for cognitive therapy as it is for other therapies.[7] One researcher looking at which factors contribute to a successful outcome in psychotherapy found that warmth and empathy on the part of the therapist could account for fully 30 percent of what helps a patient improve.[8]

It seems likely that if we can better understand the mechanism through which caring and acceptance promote psychological healing then we can better utilize it in the psychological change process. This turns out to be true.

When we feel caring warmth or when we share that with others, the brain releases a powerful array of hormones that we experience as pleasurable.[9] These hormones seem to be powerful contributors toward both physical and psychological health.[10] They include opiods, oxytocin, endorphins, and morphine-like substances. Although neurobiological researchers, like people in general, make distinctions between different kinds of warmth or love (such as parent-child, romantic, between intimate friends) research suggests that all of these types of caring involve the release of powerful, pleasure-inducing hormones. (The power of these hormones is underlined by the fact that many of them have been shown to play a significant role in addictions.)[11] Furthermore, studies have shown that the same hormones that are activated by warm interactions serve to modulate the intensity of emotional responses.[12]

Research suggests that those interactions which produce warm, loving feelings that lead to psychological well-being include two qualities shown to be especially important in order for healing to occur.[13] They are accurate empathy and a genuine, warm, caring concern for the other person—or for oneself, we might assume. Sincere empathy for a person's distress, followed by an expression of genuine, compassionate concern is likely to induce in them, at least temporarily, an experience of well-being. Interacting with someone in this manner is often referred to as "unconditional positive regard" by researchers.[14] In the third step, we want to offer ourselves an experience that includes both empathy and warm, caring concern.

At first glance, this may seem like a tall order—we need to offer sincere empathy to our self and arouse a warm feeling of being cared about, and we need to do this while a script is activated! How can a person spontaneously call up positive feelings while experiencing an amygdala script that is characterized by negative feelings? It turns out this is not so difficult as it might seem. In fact, psychotherapists have, for decades, been teaching their patients a means by which they can do this.

____Practicing Mindful Personification

The method that we have in mind is personification.[15] Fritz Perls probably did the most to popularize personification, making it a mainstay of gestalt therapy.[16] Personification, when used as a tool for personal growth, means consciously projecting our own feelings/imagery/self-beliefs upon some imagined entity.

At the beginning of this chapter, we introduced Steve, Martha, Mike, and Geri to portray examples of practicing the third step. In each of these cases, they had learned to identify with their former self who existed during those times when the script was learned. These people used their former self to personify their script. More specifically, the seed images that we discovered in the last chapter include a memory of a younger self, and the feelings and beliefs that they were harboring when the targeted script was being encoded. By serving as a representation of the feelings, imagery, and beliefs associated with a script, that former self personifies the target script. It is easy to warmly empathize with that former self, while remaining mindfully separate from him or her. Through personification we can both activate a script and work with it, yet not become subsumed by its emotions and imagery.

It may seem that personifying a script as the person you were at the time of the seed image will be difficult. It is actually quite natural. Personification is a natural human tendency. People have been personifying their psychological difficulties since prehistoric times. Spirits, evil, or punishing deities, sorcerers with whom we do battle to relieve evil spells, and the visitations of the dead are but a few of the images we have used to address psychological problems. The ancient Greeks provide us with some of the most beloved examples of personification: Eros personifying the trials and tribulations of romantic love, Aphrodite personifying the intensities of feminine passions, and Zeus personifying both the useful and the destructive aspects of our desire for power are a few well-known examples. The practice of assigning psychological phenomena to entities such as deities, spirits, ancestors, and demons is so commonplace that James Hillman, the great Jungian scholar and the originator of "archetypal psychology," refers to personification as a quintessential human practice for dealing with psychological phenomena.[17]

When we use personification in the third step, we engage the younger self who was present at the time of a seed image. We begin by empathizing with that younger self, noting the warm feeling that arises in our body as that person feels no longer so alone with the pain. Neurobiologically, we are activating the hormones that are associated with warm empathy. This empathy helps extend compassion and likability to a previously rejected part of our self. Martha appears to her former self and acknowledges how that former self felt and also how she perceived herself as bad and evil; Mike, in returning to his former self, acknowledges how he had assumed he was terribly flawed and deserving of a beating; Stephen's former self learned to think of himself as useless and unworthy of love; and Geri's former self was certain that her sister was mistreating her because Geri was unlikable.

Conditioning feelings of warmth and empathy to a script follows naturally when we allow ourselves to be touched by the hurt of our former self. You can imagine yourself being in the seed image with your former self, and you have a natural empathy for that person because you are attuned to the feeling component of your script. This allows you to tell your younger self that you understand exactly what it feels like. You know how badly you have been hurt, and you care. You then imagine that your younger self takes this in, experiences what it feels like when someone truly understands, and no longer feels so alone with the hurt.

When we introduce our present adult self into the seed image, we are in just the right place to reassure our former self, because we know exactly what kind of reassurance she or he needs. Maybe we say, "You're not stupid. That's not the real problem. Your parents are critical. That's the real problem." Or, it may be that our former self needs to hear that she or he is good-hearted and fundamentally loving. This is true even if someone is very upset and your former self cannot help them; or even if you just made a big mistake and hurt someone. In that case you can be clear and honest in acknowledging the mistake but you still have compassion for how badly it feels and you want to reassure your younger self that basically she or he is all right.

Francine Shapiro's EMDR offers a suggestion that a number of clients have found very helpful in the third step: the younger self imagines a self-affirming message written upon the wall of the room or in the sky. Perhaps the message says, "You're okay. There is nothing fundamentally wrong with you," or perhaps, "You will someday be empowered to protect yourself from being abused."

Neurobiology points to the best timing for when the brain might be most amenable to such a statement. Research has shown that when a person experiences a moderate level of psychological distress, he or she is most open to reassurance from others. If the distress is too intense, then the hormones associated with warmth and well-being are less likely to be activated, but a moderate amount of distress actually increases the probability of such hormones being released.[18] We have already introduced in the previous chapters methods to activate some, but not too much, emotional intensity: step 1 serves to activate the emotion component of a script, while the exposure exercise in chapter 5 serves to reduce its intensity to a moderate level. This paves the way for step 3—the conditioning of experiences of warm empathy and well-being to the script.

Creatively Using Seed Images and
_____Personifications in the Third Step

We are using seed images and the personification of our younger self as metaphors. Because our younger self is only a metaphor, we do not need to be concerned about whether or not our personifications are realistic. By imagining our present self appearing at the time of the seed image, we are already taking great liberties with the reality of that time. When we imagine that our former self experiences our empathetic caring and physically feels what one feels when one is met with warm empathy, we do not care if such an exchange would be a realistic fit for the circumstances of that time.

For example, if you think that you will find it difficult to imagine your younger self feeling the warm empathy that you are trying to offer, practice by imagining someone else with you in the seed situation—perhaps a kind, caring person whom you have known before. Or you might use the image of an especially compassionate religious figure; anything that works to create a feeling of warmth and empathy is fine. While doing the third step, one person included in her image the mother of her childhood friend—a woman who had been especially warm and caring toward her. Another person imagined the Dalai Lama to be present and comforting, and yet another would simply think of his wife, and this would be enough to engender an experience of being the object of warm, empathetic understanding. Many people have called upon images of their grandparents. Once you have an experience of being understood and appreciated, notice the sensations in your body, even very subtle ones that connote these feelings. This helps you to become more familiar with warm, caring feelings and makes it easier to evoke the feeling in the future.

In addition to the reinforcement contributed by warmth and empathy, there is another very powerful reinforcer that we have included in step 3. If we think of calling upon feelings of warmth and empathy as the first part of step 3—step 3a—then we might call the next part step 3b. Step 3b entails calling upon feelings associated with a basic sense of well-being. Evidence suggests that the hormones associated with a generic sense of well-being are different from those associated with warmth and caring. Whereas pleasurable, warm, caring feelings, derived from an experience of empathy, have been shown to be especially strongly correlated to the release of oxytocin in the brain, it is the release of dopamine and opiod hormones that induce more generic feelings of well-being.[19] These latter hormones provide powerful reinforcement for the conditioning of new responses that replace old habitual behaviors.[20] In step 3b the response that we are conditioning is the self-affirming and reassuring belief that we have formulated to replace the dysfunctional belief that we learned in the seed image.

It is okay to be very creative in the way in which you evoke feelings of well-being. It makes little difference if you were capable of such an experience at that time. You are simply using the seed image as a means of conditioning a new response. Ways in which people evoke an experience of well-being include remembering what it felt like when something good happened to them, or remembering happy times, maybe with friends, family members, lovers, or pets. People with whom I have worked on the third step have surpassed any possible efforts of mine in the creative ways they have found to accomplish a feeling of well-being. One man remembered a favorite character from a movie—someone who was feeling very good about himself. This person then asked himself "What would that feel like, to feel that good about oneself? What kind of sensations would that person or character be feeling in his body?" Then he imagined those sensations in his own body. Many people remember the good feelings that they know through their yoga and meditation practices, and others through prayer. Others recall what it felt like when they received an award. Some think of playing with their children or their pets, noticing the good feelings that arise in their bodies as they access these memories.

Once you have found a way to evoke a feeling of well-being, practice noticing the subtle sensations in your body that you associate with that feeling. Noting the physical sensations associated with well-being aids you in being able to more readily engender that feeling in the future.

How To Practice the Third Step

To set the stage for the third step, we tune into our former self in the seed image and ask our self: "What seemed true about me at that time?" "What was the dysfunctional belief that I was taking in about myself at the time of the seed image?" Once that has been answered (the answer may change a bit as you practice with a script), then ask yourself what would be a much better, opposite belief about yourself—one that would counter the old scripted belief and that would promote a feeling of psychological well-being. For example, when a person is working with a seed image where he or she was abused or felt helpless, the new belief will probably include, "Someday you will be empowered to protect yourself and be in charge of what happens to your body." Often the new belief includes, "You are very loveable." And always a new belief at least implicitly includes a statement to the effect that "You are fundamentally fine."

Despite the rather lengthy explanation of the third step, the actual practice of it is quite easy. To experience it do the following:

- ☑ Pick a script and practice the first two steps.
- ☑ Imagine your present-day self magically appearing in the "seed image."
- ☑ Imagine your present-day self putting your arm around your former self or otherwise engaging her in a caring, empathetic manner. (Here you may imagine other caring people or figures present as we suggested above.)
- ☑ Imagine telling your former self that you completely understand her emotions. (In fact, you are feeling the same emotions yourself.)
- ☑ Imagine that your former self—still in the seed image—lets their body relax in the way that a child does when she realizes that finally someone truly understands.
- ☑ Now tell your former self the new belief about yourself and imagine her taking it in and feeling the truth in it. (As we mentioned above, this is done with the backdrop of a feeling of well-being, garnered in any manner that works.)
- ☑ Hold that good feeling for about fifteen seconds.
- ☑ Return to the present and tune into the sensations of the world around you as if you were a movie director noting the sounds, feelings, smells, and other sensations that exist in the world.

We have now explored each of the three steps that constitute the practice for mastering amygdala scripts. In the next chapter, we will put these steps together, refining the actual practice format for these steps. The modifications found in the next chapter will serve to make utilizing the steps more straightforward and powerful. All of the three steps have important things in common. Each step represents a means of cultivating a warm, accepting, and compassionate attitude toward our mind as we engage some of the most problematic aspects of it. Each of the steps also allows us to realize that we can afford to engage with our minds fearlessly. Finally, we can accomplish these things because the three steps entail a method that allows us to observe the emotions/imagery/beliefs that arise in our mind from the standpoint of an observer. In cultivating the observer, we create psychological space from which we can watch what arises, neither hiding from the pain nor numbing ourselves against it, but establishing a healthy perspective on it. In this way, we acknowledge our scripts but do not become entangled or buried by them.

CHAPTER 7

Putting It All Together

THE THREE-STEP PRACTICE

This chapter includes little theory or research. We are ready to turn directly to practicing the three steps.

Review of the Three-Step Practice

STEP 1

IDENTIFY AND NAME THE
EMOTION COMPONENT OF A SCRIPT

1 Note what upsets you at different points of the day and pick one of these situations for practice. (You may use the same situation over and over again.)

2 Tune into the emotion associated with that situation and (if you can) associate it with subtle sensations in your body.

3 Distinguish between those emotions that are old and scripted, those emotions that are situational, and those emotions that are empathetic—if necessary refer to chapter 4. (Once accomplished, this step can be eliminated for a particular script.)

4 While noticing the physical sensations that are associated with a scripted emotion, say: "There (in my body) is that emotion."

STEP 2

FIND THE SEED IMAGE COMPONENT
AND VERBALLY ACKNOWLEDGE IT

1 Identify a seed image using the exercise ("Identifying a Seed Image") described in chapter 5. (Once you have identified a seed image, you no longer need to do this step.)

2   If necessary, use the exposure exercise described in chapter 5 or the relaxation exercise described in appendix A to reduce the intensity of the emotions associated with that image.

3   Call up that seed-image. Note that the emotion component that you are feeling is also what you were feeling at that time.

4   Say to yourself, "That's where I learned this feeling."

STEP 3
CHANGE THE BELIEF COMPONENT

1   As you feel the emotions that are associated with the seed image, ask yourself, "What were the dysfunctional beliefs about myself that I was learning at the time of the seed image?" Followed with "What would be an opposite and helpful belief to know about myself?" (Once you have done this it may no longer need to be repeated.)

2   Imagine your present-day self magically appearing in the "seed image."

3   Imagine your present-day self putting your arm around your former self or otherwise engaging him or her in a caring, empathetic manner.

4   Imagine telling your former self that you completely understand how he or she feels.

5   Imagine that your former self lets his or her body relax, feeling the warmth of acceptance in the way that a child does when realizing that finally someone truly understands and he or she is no longer alone with a bad feeling.

6   Use a memory, image, or some other method to evoke a feeling of well-being. As we suggest in chapter 6, you may be very creative in the ways that you evoke this feeling. While still experiencing that feeling in your body, tell your former self the new belief that you discovered in step 3.1 above. Imagine your former self accepting the truth in this new belief and hold the experience of well-being for fifteen seconds or longer. (If the emotional intensity is so great that this is impossible, go to chapter 5 and practice the "Exposure Tool" exercise.)

7   Return to the present and tune into the sensations of the world around you as if you were a movie director noting the sounds, feelings, smells, and other sensations that exist in the world.

___The Three Steps Condensed

After you have practiced the three steps a few times you will only need to remember a much more condensed version:

STEP 1 "There's that feeling" (as you notice in your body the feeling component of a script).

STEP 2 "There's where I learned it" (as you connect your feelings to a seed image when you had the same feeling and were learning the script).

STEP 3 "I completely understand how you are feeling and I'm here to tell you that you are fine. There is nothing fundamentally wrong with you" (as you imagine empathizing with and reassuring your former self who existed at the time of the seed image).

___How to Practice the Three Steps

Your mastery of a script will be accomplished much more quickly if you practice the three steps during sessions that you have planned for just that purpose. What I recommend is that you set aside twenty minutes a day for a practice session. If you do not practice every day or if you want to practice more than that, that's fine. The more you practice, the more quickly your mastery will develop.

Begin your practice session by assuming a good posture—align all your vertebrae such that you are as perfectly upright as is possible. Let your spine effortlessly support your head and shoulders, allowing your body to relax. Research shows that good posture encourages a positive and self-assured attitude.[1] The U.S. Marine Corps knows about this and so, to promote in each cadet a sense of fearless determination, they teach recruits to maintain a very upright posture. For our purposes, good posture promotes an attitude of confidence—"I am definitely going to master this script."

During each practice session, repeat the three steps as they are described above seven times. Then do another seven repetitions of the three steps, but for the second set do step 3 differently. Instead of imagining yourself magically appearing at the time of the seed image to empathize and reassure your former self, do the same process but use a recent situation—one that activates the amygdala script that you are targeting. For example, you might imagine yourself in an evocative situation at work, or with a spouse, or friend, or enemy—any recent (or foreseeable situation) that gives rise to the particular script you are working on. Then imagine feeling a warm empathy for yourself in that situation. Call up a feeling of well-being while still imagining yourself in that situation and remind yourself of a self-affirming truth such as "I'm fundamentally okay," "I'm empowered to protect myself," "My feelings, thoughts, and desires are valid even though I may choose to not communicate them," or some other empowering and self-affirming belief that displaces the dysfunctional belief you learned at the time of the seed image.

A note about seed images: if in steps 2 and 3 you find that your attention is being drawn to a seed image that is different from the one you first selected, use the image that your mind seems naturally inclined to focus on. It is probably more intense and more salient. Similarly, when after working with a seed image for awhile, you discover that that seed image has lost its emotional intensity, you might want to move on to another one that continues to arouse strong emotions. Scripts and people are unique, so feel free to experiment with choosing which seed image to focus upon. For example, some people find that it is better to focus upon a less intensely emotional seed image first, to gain confidence before moving on to a more evocative seed image.

That's it! Learning to master an amygdala script is as simple as that—spending fifteen or twenty minutes each day practicing the three steps—two sets of seven repetitions. As an adjunct to their practice sessions, many people find it helpful to practice these exercises while they are waiting in line, taking a shower, or even driving. When your script spontaneously comes to mind, you can use these occasions to spend the twenty or thirty seconds required to briefly practice the three steps. If you practice regularly, sometimes even for just a couple of weeks, you will probably find that when your target script is activated by a life situation, you can quickly recognize it, do the three steps, and set the script aside, clearing your mind of the usual scripted distortions—all in about twenty seconds. Especially at first, you may have to repeat the Three-Step Process a few times to maintain script-free clarity in an activating situation.

As you are mastering one script, you will probably notice other scripts. These scripts may be having a significant impact on your life or they may be simply bothersome. Most people eventually discover three or four scripts that are problematic. Some scripts remain latent, or are subtle in their effects, so over a lifetime you can expect to discover others. In learning how to master your first script, you have learned the algorithm for the mastery of all of your other scripts. An occasional review of the last pages of chapters 4, 5, and 6 sometimes helps in addressing a newly discovered script. Whenever we engage and master a script, we overcome obstacles that have limited us, obscuring some of the rich potential that life offers. Understanding this, you can cultivate an attitude so that every time you notice a script—either a previously known one or one that is newly discovered—you will welcome it as an opportunity to further free yourself from unconscious constraints.

Once You Have Mastered a Script: Post-mastery Mop-up

As they master a script, people find that situations previously encumbered with the script are now more open. Often this uncovers the need to learn new skills. Jack depicts a typical example of this. He used to get anxious in most social situations and withdraw. Later, after practicing the three steps for a while, his anxiety was greatly reduced. However, because he had habitually isolated himself, he now was not quite sure how to handle social situations. So he had to learn how to be an effective listener; how to discriminate between when his desire to speak would add to the conversation and when it would not; and how to express and assert his feelings when it was appropriate.

Alice offers another example of learning new skills as a result of mastering a script. Before mastering her script, she had found winters to be inevitably depressing. As the script that underlay her depression became tamed, she had to learn how to spend time outside in wintry environments and how to tune into the beauty of winter. Betty's story depicts another example of having to learn new skills after the mastery of a script. Previously, when her partner was critical of her, she would angrily escalate. Now her reaction to her partner's criticism was much less intense, but she had to learn how to heed some of the criticisms graciously, skillfully question other criticism, and give constructive feedback to her partner in the face of yet other criticism.

Kent had traced his script back many years to the death of his younger brother. As he mastered the script's guilt feelings and self-recriminations, he found himself grieving and so he learned how to proceed through the different phases of grief. There are many good self-help books that can aid in post-script mastery—books on communication skills and assertiveness, the grieving process, mindfulness, positive psychology, and relationship-building, to name a few. A few sessions with a good psychotherapist can also be a big help in learning new skills, especially if you feel stuck.

Metascripts: Sometimes Getting Help Is the Best Self-Help

Fortunately, for most people, scripts present themselves in a more episodic than chronic pattern—certain situations activate scripts, but more often scripts are not activated, or if activated, they are only mildly problematic. Although a script may influence how we see ourselves ("I have a temper," or "I go through periods of depression," or "I struggle with anxiety"), for most people it does not permeate their lives. In fact, most of us find it fairly easy to identify the particular situations and life events that activate painful and confusing scripts. Generally these scripts are easily addressed through the Three-Step Practice.

However, for some of us the effects of our scripts are not so circumscribed. In this case, we might think of our scripts as "metascripts." With a metascript, it feels as if the script is activated almost continuously. This results in the belief component being experienced as part of a person's identity. Metascripts are almost always derived from histories in which traumatic or very hurtful situations occurred repeatedly, or in which the seed situation was very extreme, such as being a victim of (or participating in) a very violent act or not saving someone from some terrible fate. If you suffer from a metascript, you should consider finding a psychotherapist.

How do we know if the script we are addressing qualifies as a metascript? We can get a pretty good idea of this by asking ourselves two questions: "How many areas of my life are seriously affected by this script?" and "How long has this script been profoundly affecting me?" If, for example, your depression strongly affects your social life, is harmful or limiting to your career, and you have been struggling with it since childhood, then a therapist's help is highly recommended. Similarly, if problems with anger, anxiety, or impulsivity have hindered many of your life goals and done so for decades, then we strongly urge you to seek out a psychotherapist to help you master your scripts. If you wish, you may try applying the Three-Step Practice to what you suspect is a metascript, and you may find it helpful. But pay special attention to the cautions in the previous chapters where we suggest getting professional help.

The same goes for substance abuse and dependencies. If you regularly use any drug, including alcohol, in such a way that it has negatively affected your relationships, caused problems with your health, caused repeated legal problems, or limited you vocationally, and your use continues to be a potential problem, then working on amygdala scripts might not help very much—it might even result in an increase in your substance abuse. Get help for your drug dependency first, and when that is under control work on the three steps.

____Some Notes about Observing Scripts

In practicing the three steps you are learning to be mindful—to practice observing what is going on in your mind by simultaneously experiencing and watching your experience. It is as if you are on a platform, separate from the flow of thoughts and emotions that arise in your mind and, at least for a moment, you are in a psychological space from which you can observe rather than get caught up in your mind's flow of mental activity. In the first two steps, this "platform" or "observer's space" allows you to identify objectively the emotion component of a script and where you learned it—"There's that feeling. There's where I learned it." While you observe the emotion and while you observe yourself in the seed image, you inhabit a neutral psychological space separate from an activated script.

When we say that observing or being mindful creates a "psychological space" we are of course speaking figuratively, but in fact that way of describing mindful observation is not far from the literal truth. In mindfulness, as it is used in the three steps, we are literally viewing the activities induced by the amygdala from the vantage point offered by other areas of the brain—we are activating one psychoneurological space (the neocortex) from which we can become aware of and modify the doings of another different psychoneurological space (those activities that are amygdala-mediated).

We have noted this process of mindful observation in slightly different language before, but what we now want to emphasize is the possibility of becoming curious about the actual experience of the psychological space itself. Increasing our attunement to the psychological space of the observer holds the potential for an even greater sense of mastery over problematic emotions and imagery. Lest this sound too weird, you can try to experiment with it in a very direct and simple way. Call up an amygdala script. Notice the emotion component as it arises in your body. Spend a moment noticing what it feels like to be the observer. When your attention to the emotion strays, again evoke the feeling component of a script and briefly note what it feels like to be an observer. Do this three or four times, briefly noting what the observer's space feels like.

Now imagine that you are standing on the side of a stream watching the water flow by. Imagine that in the water are the emotions, thoughts, and imagery that arise in your mind. You are simply standing and watching them go by. Ask yourself, "What is it like here where I am standing?" You will probably recognize it as a calm place. Some people report that it seems familiar—as if they have been here many times before but never really taken note of it. Usually people experience the place of the observer as pleasant and peaceful.

Most people find that they can identify with that space for only a few moments before they become distracted by other thoughts and feelings. Others might find that they can hang out there a bit longer. In any case, one of the nice things about this practice is that every time you notice a script you have the possibility of enjoying for a moment the place from which you are observing that script. In Buddhist meditation, mindfulness practices serve to aid the practitioner in observing how his or her mind habitually creates and recreates reality by constantly extracting information from the mind's flow of thoughts and feelings. Equanimity results from increasingly identifying one's mind with the observer's space. Frequently visiting the observer's space and consciously noting it might provide glimpses for us of the interaction that occurs between our minds and our experience of reality. Even this degree of mindfulness seems to activate small dosages of the pleasant experiences associated with meditation practice.[2]

Although practicing this may not afford us the profundity of those states of compassion and equanimity wherein the masters of Buddhist meditation dwell, such a practice can nevertheless take us a long way toward promoting psychological health and a sense of well-being in our life.

An aspect of what we have just noted requires emphasis. We are not leaving the rich, feeling world of scripts so that we may instead reside in a world in which painful feelings are avoided. In terms of the observer metaphor described above, it is unwise to stand on one bank of the river and look across at the activities of an amygdala script on the opposite bank in an attempt to divorce one's self from emotional reactions. With mindfulness, we definitely cross over to a calmer more pleasant place than one would find when subsumed by a script, but we are not trying to rid ourselves of emotional reactions—even those brought on by a script.

Through the virtues of self-awareness, we build a bridge from each script to a place of mindfulness. We learn to master scripts—to tame them so that they are no longer a problem, but we do not have as our goal the elimination of scripts. It is a costly gambit to attempt to dispense with messy and painful feelings. If we try to finesse our relationships with scripts once and for all by ensconcing ourselves in a world of conceptual realism or bliss, divorced from negative emotions and from our bodies, we are in danger of losing ourselves into the sterile world of the intellectual observer, or the superficially blissful acolyte. Both of these paths lead to an isolated disconnectedness. Although the most satisfying psychological stance avoids getting entangled and attached to the messy world of emotions, it does not disallow such feelings. Instead it gives us confidence to engage in relationships and other nonrational, feeling-laden arenas of human experience—the very arenas that provide the basis for the colorful adventure of a life well spent.

Our ability to observe our emotions calmly is empowered and useful precisely because we have established an intimate relationship with the emotions that we are observing. With enough practice we can come to see our scripts as allies, not enemies—the passing experience of a bittersweet memory from our past. The bitter part of our memory honors the memory's painful past; the sweet part of the memory includes warm, positive regard for ourselves and a sense of well-being in the present. These positive feelings are also induced and enhanced by the mindfulness aspect of the Three-Step Practice.[3]

CHAPTER 8

Therapist to Therapist

AMYGDALA SCRIPTS
AND THE THREE-STEP PRACTICE
IN THERAPY

We mentioned in the introduction that amygdala script theory has enhanced the therapeutic work of a variety of therapists. This chapter is written to make the benefits of this theory and the Three-Step Practice available to a broader audience of therapists.

Amygdala Script Theory and the
Three-Step Practice as Therapeutic Tools

Amygdala script theory is most usefully conceptualized as a tool that is easily applied in a variety of different theoretical orientations. Psychotherapists seem best served when clinical innovations are cast as tools of the trade. If, instead, new knowledge becomes ensconced in a particular school of psychotherapy, there is less likelihood that it will aid therapists with different theoretical orientations. The natural tendency to defend and advocate for one's school of psychotherapy reduces curiosity about the tools that are being applied in seemingly rival schools.

This is unfortunate, because much can be gained by sharing therapeutic tools across the boundaries of psychotherapeutic orientations. For example, a psychoanalyst's effective use of interpretation might be enhanced by knowledge of cognitive therapy's well-researched concept of cognitive restructuring. Likewise, the cognitive therapist's effectiveness might benefit from the subtle nuances of the therapist-client relationship that psychoanalysts have carefully explored and articulated. Both the psychoanalyst and the cognitive therapist might benefit when they ponder the possible applications of Perl's powerful use of personification and empty chair dialogues. And all psychotherapists can arguably benefit from a Jungian explication of the role that archetypes play in the daily life of our clients. We also all benefit from exploring in depth the discoveries of attachment theory and ego development, insights provided by the schools of ego psychology. Yet another example of cross-fertilization occurs when we think to frame our clients' contrarian behaviors as attempts to resolve their problems, per Weiss and Sampson's control/mastery theory, promoting in us a more compassionate understanding of why clients seem sometimes to make their lives (and therapy) so difficult. Perhaps the most useful way to view psychotherapeutic orientations is as storage sheds for tools that can be commonly applied in a wide range of therapeutic situations.

With this in mind, it is my hope that the Three-Step Practice and the theory of amygdala scripts will be understood to enhance, not compete with, existing forms of therapy. All effective therapists have an array of proven interventions that they reach for when needed. Along with those mentioned above, we might imagine a therapist's tool bench to include an effective relaxation training exercise, a few cognitive therapy tools, a well-honed knowledge of the grieving process, the ability to teach communication and assertiveness skills, a protocol for exposure therapy, the insight provided by ego-psychology into the underpinnings of very chronic problems, DiClemente and Prochaska's motivational model, and Linehan's application of mindfulness in managing intense emotions, to name a few. We can now add another very good tool to that bench—amygdala script theory and its offspring, the Three-Step Practice.

_____The Therapeutic Contributions of Amygdala Script Theory

Below is a list of some of the more important contributions that amygdala script theory promises for psychotherapy. The list is followed by a brief discussion of each concept.

1   Amygdala script theory offers a means to strongly promote therapeutic alliances by offering a simple, easy-to-understand model that explains the origins for most psychological problems while providing a scientific basis for therapeutic interventions, including the Three-Step Practice.

2   Most psychological problems are nonpathological—the result of natural processes that the brain was designed to enact—a tenet that is

reassuring and empowering to our clients. In contrast to the pathological and classificatory diagnostic system that, despite its proven
flaws, remains prevalent today, amygdala script theory promotes a
readily applicable, functional approach for describing psychological
problems.

3 Fundamental psychological change occurs by conditioning that
requires repeated practice while a script is activated—an insight that
can be applied to many types of psychological intervention.

4 Mindfulness plays a powerful role in most psychological change
processes.

5 Gestalt therapy's personification tool can be utilized to induce oxytocin and dopamine as powerful reinforcers for psychological change
mechanisms.

6 Amygdala script theory offers a potent enhancement to exposure
therapy.

CONCEPT 1
AMYGDALA SCRIPT THEORY PROVIDES A MEANS
TO STRONGLY PROMOTE THERAPEUTIC ALLIANCE

Amygdala script theory offers a simple, easy-to-grasp conceptualization
for most clients' problems and suggests a way to proceed to address those
problems. This is fertile ground for the first task in therapy—the creation
of a strong therapeutic alliance between client and therapist.[1] Bordin
summarizes three elements of the therapeutic alliance that have been
shown to be vitally important for successful treatment outcome. These
are: mutually accepted goals for therapy, a valid and mutually agreed
upon theoretical understanding of the problem, and a shared perception
of what is the best means for accomplishing treatment goals.[2] The ways
in which amygdala-mediated memories work, their roles in our everyday
lives, and the manner in which the Three-Step Practice can lead to mastery of these problems, offer a simple, appealing, scientifically based way
for client and therapist to collaboratively agree on these three: goals, theory, and therapeutic process.

In a first meeting with clients, if they are asked to describe their com-
plaints in concrete form, using specific instances as examples, they will
almost inevitably end up describing either the image component of a
script or the feeling component, often both. Below is an example of a typical client/therapist dialogue that portrays this.

The therapist has been exploring with a client what brought her into therapy. In response to the therapist's questions, the client has just finished describing specific situations that are most difficult for her.

> THERAPIST: So, am I understanding this correctly—you're describing the situations that seem to especially activate your problem?
>
> CLIENT: Yes.
>
> THERAPIST: In the situations that you've just described, you felt *[describes the bad feeling: anxious, self-critical, fearful, hopeless, irritated, confused, helpless, or otherwise problematic emotions, using the client's words]*, and what seemed true in that moment was that [describes the image component, again using client's words]. Am I getting this right?
>
> CLIENT: Yes. You seem to be getting it very well.
>
> THERAPIST: Let's try an experiment. As you imagine *[therapist reiterates the situation that the client has just described]*, can you call up those same feelings again? *[Client is almost always able to do this.]* Now describe what those feelings are like in your body. In other words, where in your body do you associate those feelings?

Most clients are able to do this. Sometimes some guidance from the therapist helps. If a client is unable to do this, the discussion proceeds but with reference to the emotion and perhaps some of the imagery of a script rather than the physical sensation that the emotion elicits.

> THERAPIST: Let's check something out. Describe a few, maybe two or three, more situations where the same problem seemed to be especially strong.

As the client does this, each situation is examined to see if a similar emotion and imagery arises. Client notes the similarities and in doing so refines her understanding of what has been causing her problems. Occasionally, an example indicates a different script. The difference in feelings and imagery for that situation is noted and set aside for future reference.

> THERAPIST: So it would seem that when this problem arises, it manifests with the same emotions in your body and similar assumptions—similar images—of what it seems like is going on in the moment. Is that right?

CLIENT: Yes, that's becoming very clear.

At this point, if the client has not already acknowledged the exaggerated or distorted quality of the image component, she is encouraged to examine how fitting the image is and what aspects of it might be misleading.

THERAPIST: I think that what we're discovering here is that at certain times in certain kinds of situations *[the problem]* gets especially activated. What is causing your problems appears to have at least two components. *[Therapist describes the emotion component and the image component, again using the patient's words.]* So far so good?

CLIENT: Yep.

THERAPIST: This is what I refer to as a script. I call it that because while it's activated, it tends to "script" our feelings and our assumptions about what is going on.

Let's check out one more thing—whether or not the feelings that come up in your body are old and familiar feelings. Imagine one of these situations that you've just described. Stay with that memory until you notice the emotions arising in your body and see if those emotions seem familiar—if it seems like those particular feelings have often come up before.

Generally, after noting the situation and the concomitant emotions, the client readily acknowledges that those particular feelings are familiar and have been a problem, usually for a long time.

Typically, at this point the therapist describes the neurobiology of amygdala scripts, how these scripts are a naturally occurring phenomenon, how they were useful in our evolution, and the underlying neurobiology that results in the script being activated without conscious awareness. By now, the client has probably mentioned a belief about herself (the belief component) that she readily recognizes as exacerbating the damaging effects of her script. If this has not occurred, the concept of the belief component is described in terms of her target script.

Covering the material described in the above dialogue will often occur over the period of a few sessions, but I continue to be surprised that the early sessions with a client culminate in a dialogue very akin to that above with more than half of the people I presently see. Clients are almost uniformly very pleased to have this new conceptual understanding of their difficulties, but they want to know something more: "Knowing what is going on that has been causing this problem is all well and good, but is there some way to help me with it?" This flows naturally to a description and rationale for the Three-Step Practice. The next step is to decide if now seems like a good time to learn this therapeutic tool. For most clients, an exploration of amygdala scripts culminates in Bodin's components of a therapeutic alliance: an agreement on treatment goals (mastery over a script's emotions, imagery, and induced behavior); a conceptual way to understand the problem (amygdala script theory); and the means to attain that goal (various therapeutic tools, usually including the Three-Step Practice).

CONCEPT 2
AMYGDALA SCRIPT THEORY FAVORS A FUNCTIONAL SYSTEM FOR DESCRIBING MOST PSYCHOLOGICAL PROBLEMS

In the last few decades, psychological difficulties have increasingly been cast as neurological diseases or brain anomalies. All too often, a DSM-IV diagnosis suggests that people who suffer psychological problems have a somewhat intractable, structural problem in their neurology. (It is beyond our focus here to delve into why, with limited scientific support, such an extreme view of psychological problems has become so prevalent.)

There *are* clinical benefits to the DSM-IV. For example, the DSM-IV classification of schizophrenia and bipolar I disorder have important treatment implications. No doubt, classificatory diagnostic systems such as the DSM-IV have aided clinicians and researchers. These types of diagnosis are prone to misunderstanding and misapplication, however. The diagnoses for psychological problems found in the DSM-IV (those that are based on sound research) are derived from factorial or cluster analyses of psychological problems.[3] Factorial analyses, when applied to psychological problems, identify symptoms that tend to arise concurrently. For example, lethargy, insomnia, and anhedonia are measurable behaviors that tend to occur together, as part of a cluster of symptoms that we refer to as depression. When a cluster of symptoms is identified and labeled, our inclination is to make two assumptions: that the cluster of symptoms has a common or closely related cause, or causes; and that these causes are pathological—they represent a biologically compromised system. This approach is appealing because it has historically been very useful. It has allowed us to specify an illness (the congestion, aches, and exhaustion associated with the flu, for example) and thereby pave the way for a search for the underlying pathological causes of that illness (the various flu viruses that attack and compromise healthy human cells). When we apply this model to psychological problems, the first assumption (the existence of a common cause or causes that underlie the cluster) is guardedly warranted, but the second assumption (that the problem is primarily caused by something wrong biologically) is doubtful, at least in regards to many common psychological problems.

At its root, a cluster analysis is simply descriptive: it describes a number of factors that cluster together and that appear to be related. Labeling a cluster suggests little or nothing about what caused that symptom cluster. To say that someone is lethargic and anhedonic because they are depressed is redundant. The term depression does not imply a cause for anything; it only suggests a cluster of symptoms, including anhedonia and lethargy. Amygdala script theory, in contrast, focuses upon the causes of psychological problems and, as we have noted before, illuminates non-pathological causes for most psychological problems. Although these problems cause pain and dysfunction, they nevertheless are readily recognizable as the result of a brain that is functioning in exactly the way it was designed to function.

Our focus on classificatory diagnoses would be of only academic interest if such diagnoses did not have the potential to harm our clients. The negative effects of pathological labels on clients have been well publicized.[4] We can assume that a client who leaves his therapist's office with the insight that his depression is the result of an amygdala script—once learned and now tameable—is more prepared to engage psychotherapy effectively than the client who has been told that his depression is the result of a biologically based brain anomaly. And the client's willingness to engage fully in a course of psychotherapy is very important to maintain long-term improvement from depression, anxiety, and obsessive-compulsive disorders.[5]

Amygdala script theory suggests that we as clinicians serve our clients best when we diagnose in terms of a functional and prescriptive focus on symptoms rather than in terms of categories. Categorical diagnoses, potentially damaging in their labeling, also have the disadvantage of offering limited direction for treatment planning. For example, two clients, Tom and Mary, both had been told that they were suffering from a "Major Depressive Episode." Indeed, they both qualified for this DSM-IV diagnosis. Tom complained that his life felt empty and meaningless, he had been overeating and gaining weight, he was oversleeping, had little energy, and trudged into my office showing obvious signs of psycho-motor retardation. He had just lost his job and was socially isolated. Mary appeared to have a normal amount of energy but she complained of frequent bouts of sadness. She was often tearful and generally anhedonic, had experienced significant weight loss, had been suffering from insomnia, reported problems with concentration, and had frequent periods wherein she felt agitated. Mary is married to a supportive husband. She was preoccupied with the death of her mother a few years earlier. In contrast to Tom's social isolation, even though she no longer enjoyed many of these social activities, Mary often went out with her husband to be with friends. Although the diagnosis is the same for both of these people, it is clear that many of the interventions useful for treating the one would not be helpful to the other—in fact could be disastrous for the other. Focusing on grief with Mary, encouraging her to supplement her diet regularly with protein drinks, and suggesting an occasional sleeping pill would be interventions potentially very helpful to Mary but disastrous for Tom.

Classificatory diagnostic systems when applied to psychological problems are also limited in their explanatory value to clients. No competent mechanic would call you to report that you have "a fuel system syndrome," or that your car is suffering from "a braking disorder." Instead you will be told that your carburetor is not getting the fuel/air mix right and needs to be rebuilt, or your front brake pads need to be replaced because they are worn out, leaving metal against metal to cause the squeaking that you have been hearing. These are functional diagnostic statements—they designate the functional basis for your car's problem and so allow you to understand the specifics that are necessary to get your car back on the road. Similarly, when Tom discovers how childhood abuse from his father left him with a script that, when activated, leaves him feeling helpless and hopeless, he knows what he needs to address. Mary discovered a very different script central to her problem. The belief component of her script revolved around a childhood assumption that if she were a loving and worthy daughter, she would be able to alleviate her mother's pain. As an adult, when this script was activated she unconsciously assumed that she was responsible for alleviating painful things that arose in the lives of her husband and their children. Often failing to accomplish this and under the influence of her activated script, she felt unworthy and unlovable—a miserable failure. Understanding her script and how it functioned led her to be conscious of this and to master it. Mastery of her script freed her to renegotiate her relationships with some family members and friends, and to develop the skills to nurture and replenish herself.

The functional understanding of their problems provided by the concept of amygdala scripts allowed both Tom and Mary to see how they could address their problems. Being classified as having "a major depressive episode" had not been helpful to either of these people. It did not serve to enlighten them as to how they might begin to address their problems, and the diagnosis had left them confused and intimidated.

A nonpathological description of our clients' problems gains further support through research that underscores important weaknesses in the validity of our present diagnostic classification system: the DSM-IV has weak empirical underpinnings and, as we have seen, the diagnostic categories overlap each other.[6]

Suggesting that pathologically driven diagnostic categories have critical weaknesses does not mean that there are no patterns of dysfunction that could well have a biological as well as a psychological basis. However, neurobiological underpinnings should not be assumed to be neuropathology. There is a strong body of research which suggests that different people have different predispositions toward certain types of problems, but it is important to emphasize that this does not imply that those tendencies are inherently pathological. Amygdala script theory converges with this growing body of research that points to robust personality variables which interact with environmental and other factors to differentiate what we have come to call "psychopathology."[7] The fact that personality traits remain stable even after successful treatment occurs, at least for depression, suggests that these factors measure *predispositions* to particular types of problems, but not *causes* of psychological problems.[8] As is suggested by amygdala script theory, the form that psychological problems take are likely to be strongly influenced by "hard-wired" traits, but the origins of those problems are usually found in historical environmental experiences (depicted in "seed images").

Our proposition that amygdala-mediated memories play a central role in the development of a great deal of psychological dysfunction does not suggest that the role of personality variables is trivial. The degree of genetic loading along a particular dimension can have important ramifications regarding prognosis, and especially treatment strategy.[9] Although personality variables are arguably not the cause of psychological pain, what is clear is that there are a variety of ways in which they influence the form that defenses against psychological pain might take. For example, a child who is especially sensitive and introverted, when subjected to being repeatedly shamed, will tend to develop a defense mechanism that includes withdrawal into self-critical brooding; whereas a child who has a more joyous, optimistic, and impulsive nature, after being subjected to the same repetitive shaming, might unconsciously opt for constant distractions and entertainment to avoid the pain, eschewing calmer environments in favor of those that offer more immediate entertainment and adventure. Eventually, such a child might have difficulties in establishing stable relationships and, if his or her genetic makeup is such that alcohol is especially effective in anesthetizing pain, there might be a risk of developing a drinking problem.

Enduring personality traits, perhaps genetically dictated, also have important ramifications regarding treatment strategy.[10] The person described above who is prone to self-critical brooding might benefit early on in his treatment by learning the Three-Step Practice to master his or her script. Our other example, a person oriented toward external stimulation and excitement, may, on the other hand, need to learn to listen more carefully to external feedback from friends, to experiment with periods of less activity, and to learn some mindfulness skills, so as to be more attuned to what arises in his or her mind when not being entertained and distracted. It may also be necessary to stop drinking and be sober for awhile before being ready to address the more painful undercurrents wrought by an amygdala script.

It has been my good fortune to participate in many ongoing case consultation groups. Over the past thirty years, these groups have included approximately forty different therapists representing more than four hundred years of combined clinical experience. The theoretical orientations have been varied, including cognitive therapy, gestalt therapy, therapies based on attachment theory, biologically based psychiatry, Buddhist psychology, control mastery theory, and others. I mention this because very little time in these groups has been spent on diagnostic labels. There is an implicit understanding that categorical diagnoses are very limited in their ability to illuminate what is going on with a client. What *does* occur in these consultations, almost inevitably, is that a consensus forms around what would be good for a particular client. Also, with a large majority of the clients who are presented, there has been little focus on pathological deficits; instead, the focus has been on a client's strengths and how strengths can be utilized to create opportunities to learn new skills that replace dysfunctional habits. Amygdala script theory, with its nonpathological underpinnings, resonates well with conceptualizations that eschew diagnostic categories in favor of useful interventions.

CONCEPT 3

FUNDAMENTAL PSYCHOLOGICAL CHANGE IS CONDITIONING THAT REQUIRES REPEATED PRACTICE WHILE THE TARGETED SCRIPT IS ACTIVE

All psychotherapy includes a great deal of repetition. Whether it be an analysand's repeated revisiting of historical insights and their application to present life; the cognitive therapist's client who repeatedly practices identifying, stopping, and replacing a dysfunctional belief; the client who practices a daily relaxation exercise; or the client learning to remember repeatedly to be assertive, there seems always to be a process of repeated practice that occurs in effective psychotherapy. We in the psychotherapy world have been slow to appreciate the significance of this. It has been left to neurobiology to show us that psychological problems are a product of the limbic system and other parts of the brain that learn through repetition, through conditioning while the targeted processes are active, rather than through the higher-level neurological processing whereby new information becomes immediately applicable.

The central role that focused, repeated practice can play in psychotherapy is perhaps the most powerful contribution of amygdala script theory to psychotherapy. Historically, most clinical approaches have encouraged clients to apply insight to a problematic pattern of emotion and behavior while it is occurring. Amygdala script homework extends this to include a series of "at home" practice sessions wherein mindfulness, insight, and cognitive change strategies are applied while an amygdala script is manifest. This serves to condition neocortical activity (fundamental to mindfulness, insight, and cognitive change practices) to moderate the limbic system's exaggerated response. When a scripted limbic system response is activated, the prefrontal cortex is inhibited in its ability to consciously test reality, making it difficult for a client to notice an active script and hard to avoid becoming entangled in a script even after it has been noticed. This is why trying to condition a new response utilizing only the hit-or-miss opportunities natural to everyday life is very inefficient. All psychotherapy methods have developed a means to mindfully acknowledge and target old habitual patterns, but virtually none have understood the importance of repeated, purposeful activation of problems in conjunction with neocortical reality testing. We believe that the Three-Step Practice is an especially succinct and direct way to accomplish this, and it is readily integrated into a variety of different psychotherapeutic approaches. The rule of thumb is to have clients activate scripts and apply the Three-Step Practice a few times during a session and then to encourage them to practice as regularly as possible between sessions.

All effective schools of psychotherapy have (unwittingly) developed their own means for potentiating limbic system–neocortical interactions, or else they would not be effective. When the principle of conditioning a neocortical response to an activated amygdala script is understood, it becomes easy to enhance psychotherapy by creating simple homework exercises that are tailored to the scripts of a particular client. In addition to the Three-Step Practice, there are many other ways in which practice sessions can be used as psychotherapy homework to markedly enhance the effectiveness of treatment. For example, a gestalt therapist might instruct a client to practice "empty chair" exercises at home, repeatedly eliciting and modifying an amygdala script; a psychoanalytic therapist could teach a client to repeatedly envision an interaction with the therapist to activate and apply insight to an activated complex; or a cognitive therapist might teach a client to recall an especially evocative situation repeatedly, each time stopping and replacing old dysfunctional beliefs with newly learned beneficial beliefs.

CONCEPT 4
MINDFULNESS PLAYS A POWERFUL ROLE IN MOST PSYCHO-LOGICAL CHANGE PROCESSES

Mindfulness, in simplest terms, is a practice whereby conscious awareness is brought to a variety of internal, generally subconscious or unconscious, states. This makes mindfulness a very straightforward and elegant means to engage the neocortex and bring conscious awareness to a wide variety of mental phenomena, activating the neocortical regions of the brain. It is a practice that is quickly rising to prominence in clinical research; a recent PsychInfo search for "mindfulness" in peer-reviewed journal abstracts from the last ten years generated 479 different articles.[11] Amygdala script theory offers additional understanding of some of the mechanisms that underlie mindfulness, especially as it is applied to working with emotions. In chapter 4, I noted neurobiological studies which demonstrate that the brain recognizes emotions by monitoring internal states of the body. Knowing this allows us to enhance mindful awareness of emotions to a greater level of psychotherapeutic effectiveness by teaching clients to associate their emotions with subtle sensations in their bodies.

Mindfulness, whether the term is used or not, marks the start of almost all beneficial psychological practices. We must be able to step out of the entanglements of a script and to observe its effects before we can effectively work with it. Teaching our clients a mindfulness practice— observing their emotions, imagery, and beliefs—can serve as a cardinal enhancement to many different therapy orientations. By itself, mindfulness has been shown to have a moderating influence upon psychological difficulties and, when integrated with additional neocortical conditioning processes such as those found in the Three-Step Practice, the effectiveness of mindfulness in mastering everyday problematic scripts is enhanced further.

CONCEPT 5

PERSONIFICATION CAN BE USED TO INDUCE OXYTOCIN, DOPAMINE, AND OPIOID RECEPTOR POTENTIATION TO ACT AS POWERFUL REINFORCERS FOR PSYCHOLOGICAL CHANGE

In chapter 7 we noted that empathy is well understood to play a central role in promoting psychological health. We also cited research which indicates that the hormones associated with warm empathy and well-being are very powerful reinforcers.It has traditionally been assumed that the healing effects derived from these experiences depend on external situations—a therapist/client interaction, for example—to activate the hormonal responses associated with empathy. In our use of personification in steps 3a and 3b, a practice is introduced for activating the beneficial experiences of empathy and well-being without the need for external triggers. A client learns to personify a younger self in a seed situation and to consciously induce feelings of warmth and well-being, in part, by imagining the present self empathizing and validating that younger self. These positive emotional experiences are used to powerfully reinforce the beneficial self-beliefs that are applied in step 3.

Inducing dopamine and oxytocin to promote psychological health can be applied in a variety of clinical contexts. For example, I observed a very depressed and suicidal client in my office suite warmly interacting with a colleague's dog. He and I were able to use this as a reference point for the sensations that such an encounter produced in his body and to acknowledge his good-heartedness. More important, this awareness catalyzed his ability to recognize his potential to relate to himself in ways that served to foster similar experiences of connectedness in his everyday encounters. He began to practice noticing—and soon he learned to cultivate—the simple, warm, and mildly pleasurable experiences that arose when he felt an empathetic connection with others. He began to recognize himself as a good-hearted person. His realization was made much more potent when he experienced in his body the sensations associated with warm connectedness. He learned that this experiential awareness enhanced his ability to develop the skills necessary to appreciate his world more fully and to learn to better connect with others. From a neurobiology standpoint, he has learned how to independently induce pleasurable hormones, probably especially oxytocin, and to associate the ensuing experience with a more accurate and more favorable understanding of who he is.

If we were to create a neurobiological bill of rights, we would begin it with the birthright to have ready access to the hormonally promoted feelings of warmth, love, and well-being. All parents would be encouraged to awaken these feelings in their babies again and again. Fortunately, it is natural for most parents to do this, for example to coo to a baby to elicit smiles and delight in that child. (For that matter, it is pretty natural for a stranger to do the same.) For most parents, it is natural to applaud and share delight with a baby many times a day—experiences epitomized when landmark accomplishments such as a first step are accomplished. Skillful parents continue to elicit these experiences with their offspring throughout their history together. The baby or child is being encouraged in his natural tendency to experience warmth and well-being—experiences probably associated with the hormonal release of oxytocin and dopamine. Especially with babies, the promulgation of this type of interaction by caretakers and children is a very important component for developing what researchers refer to as healthy attachment.[12] Healthy attachment has been shown to have far-reaching and fundamental importance regarding psychological health. When we teach our clients practices whereby they can elicit for themselves feelings of warmth and well-being, we are teaching them practices that cultivate basic psychological strengths.

CONCEPT 6

AMYGDALA SCRIPT THEORY OFFERS A POTENT
ENHANCEMENT TO EXPOSURE THERAPY

It seems safe to say that Edna Foa has done more to contribute to our understanding of exposure procedures than any other scientist.[13] In 1986 she began to suggest that much of the well-documented efficacy of exposure therapy in the treatment of Post-traumatic Stress Disorder is the result of a cognitive restructuring process that is inherent in exposure techniques.[14] Foa noted that after a period of prolonged emotional exposure, clients' beliefs about themselves were changed—that is, they underwent a cognitive restructuring process. Since exposure therapy alone appeared to induce cognitive restructuring, Foa wondered if the cognitive restructuring process that is typically used in conjunction with exposure therapy might be redundant. Exploring this question in 2004 and 2005, she found that, in fact, prolonged exposure techniques were so effective in allaying dysfunctional beliefs that adding conventional cognitive restructuring treatment interventions did not enhance the efficacy of the exposure treatment.[15]

Amygdala script theory offers a very useful explanation for the mechanism through which exposure therapy results in cognitive restructuring. In exposure therapy, a client evokes the feelings and imagery of an especially painful learning situation; in amygdala script theory these are the emotion component of a script and seed image. Clients undergoing prolonged exposure maintain awareness of the emotions and imagery associated with that seed situation. Presumably, during this process the client is also aware of experiencing a memory. This involves maintaining an activated amygdala-mediated memory and conditioning a neocortical response (the conscious awareness that the experience is indeed a memory) to that memory. In simple terms, the client is repeatedly experiencing an emotion memory while consciously associating that memory with its historical origins. In effect, clients are practicing reminding themselves that what they are experiencing is "only a memory," one that has little relevance to their present reality.

What then would the effect be, if in an exposure therapy exercise, while an old painful memory is activated, clients are directed to remind themselves more explicitly that what they are experiencing is an old memory? In this experiment, the client would continue to recall a traumatic memory and its concomitant emotions while the therapist reminds her that she is experiencing an emotion memory. She can also practice this on her own by calling up a seed image, becoming mindful of the feelings in her body associated with that image, and repeatedly reminding herself that what she is experiencing is an emotion memory:

"These are feelings that I learned a long time ago."

"It is a feeling memory that has no real relevance to my present day reality."

"My body is manifesting an emotion that is a memory of what happened back then; it doesn't have to do with now."

"The emotions that I'm noticing as I continue to remember what happened back then are what it felt like then, not what is going on now."

And so on.

In the 1980s I began to use prolonged exposure with PTSD clients; by the 1990s, encouraged by its success and the effectiveness of the exposure techniques inherent in EMDR, (techniques that suggested broader applications), I discovered the usefulness of applying exposure protocols to a greater variety of problems. In recent years, inspired by the research published by Edna Foa and her colleagues, I began teaching my clients to practice the simple "self-talk" described above—reminding themselves that they are simply experiencing an emotion memory—as part of the exposure therapy protocol. I have been very impressed with the result. The time for their emotional intensity to subside to a level of "two or three" has become markedly truncated, usually about twenty minutes, and the subsequent reduced emotional intensity has quickly generalized to deintensify emotional reactions to triggering situations in the daily lives of these clients.[16]

____The Psychotherapeutic Basics

No discussion of the application of a psychotherapeutic tool would be complete if we did not at least mention the basics. And the basics of psychotherapy seem to revolve around the therapist/client relationship. In particular, central to success in psychotherapy is the strength of the therapeutic alliance and how diligent we are in maintaining an empathetic connection with our clients. As a therapist, it is important to keep in mind that when a client is discussing something that is painful or confusing, if we have to choose between empathizing with that client, telling them about a really terrific therapeutic tool, or offering an important insight, we should practice erring on the side of choosing empathy.[17]

In chapter 6, I mentioned the importance of beginning therapy with three factors in place: client-therapist agreement on goals, theory, and means. What I want to emphasize now is that these three factors continue to play a crucial role throughout the therapeutic process. It is centrally important to continue to monitor the status of our alliance with our clients. An effective therapist remains alert to the possibility that the interest of the client is waning and, in exploring this, keeps in mind the status of mutually agreed-upon goals, theories, and means. As two researchers state: "the concept of the alliance 'highlights the fact that at a fundamental level the client's ability to trust, hope, and have faith in the therapist's ability to help, always plays a central role in the change process.'"[18]

Although a strong therapeutic alliance promotes but does not always create a positive treatment outcome, left unrecognized or unrepaired, a ruptured therapeutic alliance will inevitably markedly reduce the likelihood of a positive treatment outcome.[19] In almost every case where, at the end of treatment, I felt that I was not as helpful as I could have been, I can in retrospect see that at least one of those components of the therapeutic alliance had become neglected. The take-home message is that even though the concept of amygdala scripts offers an intuitively and logically appealing conceptualization of a client's difficulties, even though that understanding often progresses naturally to the Three-Step Practice, and even when my client has enthusiastically engaged both my conceptualization and method for working with his problem, I still need to remain attentive. I need to ensure that at any given juncture my client and I are on the same page, that we are both operating with the same sense of what the problem and goal are, and what the best way is to proceed.

When, as therapists, we have lost our resonance with a client, it behoves us to be willing to set aside what we consider an insightful understanding or helpful direction in deference to honoring our client's understanding of the situation. It behoves us to assign special weight to the possibility that a client's impulse to proceed in a different direction may have advantages that we have not yet understood.

Clients with Metascripts

Sometimes a client is experiencing difficulties that can be seen as arising from what has been traditionally referred to as a personality disorder (in DSM-IV parlance, "an Axis II diagnosis"). In chapter 7 we referred to these types of problems as the result of metascripts—very long-standing scripts that profoundly affect the quality of a person's relationships and that are rooted in especially painful childhoods. When these clients seek help to change the pervasive effects of chronic problems (as compared to those who seek help for more circumscribed problems), psychotherapy usually entails a much longer process, often years.

Nevertheless, these people can be the most satisfying of clients. They can be hardworking, very appreciative of what they accomplish, and one cannot help but be impressed by their courage and stamina. At the end of therapy with a client who has been severely disabled by metascripts, if you review your years together you will remember a rich array of experiences that you and the client shared, some touching, others worrisome, and many memories of being mutually challenged. In other words, it is a real adventure, the type of adventure not often afforded by most types of work. Furthermore, these clients often offer the greatest opportunities for learning that a therapist ever encounters. Missteps are highlighted because the ramifications of what might be ignored or forgiven by another client tend to be magnified in the reactions of a client with an active metascript. These clients are also those who are most likely to provide you with straight, undiluted, critical feedback.

Clients who have numerous and intense early traumas not only offer you an intense, not-so-forgiving laboratory for honing your therapeutic skills; what is more interesting, they confront you with the possibility of learning about the subtleties that are activated in your own psyche. Clients with metascripts can be uncannily unerring in their ability to home in on *your* scripts, including those more subtle scripts that you have found easy to ignore. This is because people who have been badly harmed learn to be masters at sensing in others, especially their therapist, the themes of their own scripts and finding ways of activating those so that they can see how the therapist handles such things. This allows them to decide whether or not you are a competent guide on whom they can depend for help.[20] Before rejecting you, usually they will repeatedly evoke negative feelings in you, letting you know at least implicitly how well you are doing in responding to them.

It is beyond the scope of this book to go into detail about working with clients who are experiencing metascripts. However, a few points seem noteworthy. With these clients, the most important scripts that arise usually arise through the transference—that is, in your relationship with your client. These scripts often represent so much pain that they are not readily mastered. Scripts become reference points: the client is learning to notice what is being activated for him and to manage that. *Managing* the emotions and imagery activated through a script is different from *mastering* a script. Management comes first and is made possible by a degree of mindfulness that allows a client to identify an activated script, reality test the imagery component, and use various tools to moderate the intensity of the emotion component. In chapter 9, we will briefly describe some strategies that are useful in this regard. Further information about emotional management can be found in the works of Marsha Linehan, a true master at this, in her books on "Dialectical Behavioral Therapy." These books include useful guidelines for the management of intense emotion.[21]

It may be years before clients with metascript problems can move from emotion management to tolerate the steps toward mastery of their scripts. The crucial variable for this progression has to do with the establishment and careful maintenance of a safe therapist/client relationship. When the client is ready, exposure therapy can help to allay the intensity of the emotions. Attempted too soon, the Three-Step Process activates more affect in the transference than is tolerable, and clients will often drop out of treatment.

People with metascripts tend to have scripts that are emotionally intense and frequently activated. Working with their scripts is further complicated by an overarching negative self-image. So, when they are ready for the Three-Step Practice, shorthand methods that emphasize a caring attitude toward oneself may be called for.

A shortened version of the Three-Step Practice that clients have found very effective begins with a personification very similar to that which we suggested in the third step: after noting the emotions associated with a script, the client identifies these emotions as belonging to a younger self who existed in the seed image, and imagines approaching that younger self with an attitude of kind compassion. The practice follows this general outline:

Becoming aware of the emotion component of a script leads the client to an image of the unhappy child he once was. He welcomes his younger self, saying in effect, "Hi there." This simple "Hi there" becomes the equivalent of step 1—"There's that feeling," and of step 2—"There's where I learned it." Then he says, "What is it that you need to hear to be comforted?" (step 3).

Dean represents an example of how this can work. It should be noted that Dean had been working on the longer, more detailed form of the Three-Step Practice before he adopted this shortened version. The shortened form is much more effective if it is introduced after a client has been working with the Three-Step Practice.

Dean had been depressed since he was a child, and although he had made a great deal of progress in mastering his depression, he was still assailed with daily bouts of anxiety:

"I feel anxious and angry and then I recognize that it has to do with a tightness in my chest. As I tune into these feelings I become aware of a part of myself that is about five years old. Other than to correct him and tell him what to do, his mother shows no interest in the five-year-old rendition of himself. I realize that this little boy needs to know that he is able to put his mother aside and not get entangled with her bad feelings. I let him know this.

"Now, he feels sad because he has no real mother. In order to be comforted, he needs to be told that it is all right to feel sad; it is okay to grieve about not having a loving mother.

"He also feels lonely. His father doesn't really understand him and he has no close friends. What he needs now is to know that I am here for him. When I practice with this script, I first notice that my five-year-old self is angry and frightened of his mother. Then I help him separate from her and grieve the loss. Then I let him know that I am a strong person who can call up a sense of well-being and I can reassure that younger self that I care and will be there for him."

Dean practiced this many times. Repeatedly approaching this painful script with a caring, confident attitude, Dean began to feel more generous and accepting of himself. He also learned that he did not have to exert so much energy to avoid his anxious feelings. Using the above personification, he could engage these feelings and tame them.

If you wish to work with these clients, never blame the client when treatment goes awry but also, even more important, practice being very forgiving of yourself and your limitations. I believe that these clients sometimes benefit through a series of treatment episodes. Complimentary insights and feedback from a number of different therapists often enhance a client's ability to trust the therapeutic process.

People who suffer from metascripts are frequently activated by relationships. Therefore, constant monitoring of the therapeutic relationship is called for. Empathy and the therapeutic alliance mentioned above are paramount to successful outcome. This creates a lot of demand on your ability as a therapist to monitor and work with your own aspect of the relationship. Knowing your limitations, being very empathetic while not taking too much responsibility for your client's pain are important aspects of this. Advice is dispensed, collaborative problem-solving often occurs, but the deeper therapy is occurring in the nurturance and the learning that arises from the therapist/client relationship.

There are many good books written by good teachers who specialize in working with clients suffering from what I am referring to as metascripts: Otto Kernberg, Geoff Young, Marsha Linehan and Christine Padesky are a few of them.[22]

The Three-Step Practice and Couples Therapy

There are various adaptations of the Three-Step Practice that contribute to a satisfying therapeutic experience. One of them is found in its application to couples therapy. I am of the school that believes that couples therapy can be powerfully aided by learning good communication skills. The ability to learn communication skills—including learning to be assertive, to use "I" statements, to practice accurate and genuine listening skills—can be catalyzed by practice with amygdala scripts. Being in a committed relationship is second only to having a child in its ability to activate scripts. This is not altogether bad news. A primary relationship offers rich and fruitful ground for couples to quickly recognize scripts and help each other in the mastery of them.

A common means by which couples are introduced to scripts occurs when they realize they are "speaking different languages." Mark and Martha complained that their disagreements would often escalate into yelling matches and end up quite mean-spirited. As they focused on what was getting activated for each of them during these fights, they learned that Martha felt like she "was fighting for her life." She was outraged that Mark seemed willing to sacrifice their marriage in the face of seemingly inconsequential issues, arousing an image that she and their children would be abandoned and out in the cold. Mark had a very different experience. As he sensed Martha's anger rising, he feared that he was going to get run over, be discounted, and marginalized. He epitomized this as an image of cowering in the corner unable to defend himself. To protect himself, he would "bring in the big guns," reacting very strongly in hopes of countering any possibility of being crushed and helpless.

Martha soon discovered that she was speaking the language of one who experiences herself on the brink of being abandoned—left alone in a cold world. Mark discovered that he was speaking in a tongue congruent with being in danger of being oppressed into a helpless state. Each of them watched as the other learned about and practiced the three steps. They also learned to recognize each other's script and to listen to and empathize with their partner's fear and rage, acknowledging what seemed true for their partner when she or he was activated. They soon became able to master their own scripts and skillfully sympathize with their partners. An important part of their mastery process was to acknowledge activated scripts to each other. Sometimes they would do this by rehearsing a fight: "This is not a real fight. This is a rehearsal where we try to get activated and then we describe what is activated for us." An important aspect of doing couples therapy is that both individuals have to agree to practice limiting their agenda during disagreements to the two goals of listening and empathizing with each other—setting aside, at least for the time being, hope of resolving an issue. Ironically, the outcome of practicing this is that issues then become much more easily resolved.

The Occasional and Mysterious Disappearance of Scripts after Utilizing the Three-Step Practice

Throughout this book we have emphasized the effect of the Three-Step Practice as a means to promote mastery when an amygdala script is activated. After practicing the three steps for a week or two, however, clients often report that the script they were targeting no longer arises. This may seem puzzling in light of the research that we cited above regarding the notoriously robust nature of amygdala-mediated emotion memories (that is, scripts).

A likely way to explain this is that rather than actually erasing a script in their practice of the three steps, the clients are conditioning a reframing of those situations that had previously served to activate the target script. They are learning to attach different meaning to those cues that have historically activated a script. This serves to defuse the tendency of a particular situation or set of similar situations to activate a script, a salubrious effect that is different from the mastery that occurs while a script is activated. The likelihood that this is occurring is supported by neurobiology and is congruent with cognitive and other therapeutic approaches having to do with reappraisal of events. Neurobiologically, it has been shown that if prior to the presentation of evocative stimuli a person foresees that stimuli and reappraises it (or "reframes" it, as is commonly said in psychotherapy), then the amygdala and emotional response are moderated.[23] From this standpoint, the Three-Step Practice offers an example of a priori reframing. This is especially evident in the second set of the three-step repetitions, wherein the first two steps do not differ from the previous set but step 3 brings the focus to evocative situations and changes dysfunctional, scripted assumptions into assumptions that are functional and more concordant with those present situations. This practice conditions a reappraisal to foreseeable, evocative situations. The effectiveness of a priori reappraisals is supported by neurobiological research that demonstrates that pre-event reappraisal of otherwise evocative stimuli inhibits the activity of the amygdala via the prefrontal cortex.[24] It is helpful to tell clients that a script that has seemingly disappeared may reappear again when future unforeseen circumstances serve to evoke it, but they now have the tools to master it should this occur.

New Perspectives: Psychological Mastery Rather Than Psychotherapeutic Cure

A few centuries ago, it was assumed that the key to leading a long and healthy life would be found in the advancement of technologies that would fix physiological pathologies. Now we understand that this is true only to a limited degree. We have indeed become healthier, and our life expectancy would be the envy of our ancestors, but much of what makes us healthy is not the result of curing illnesses but rather the result of healthy practices. We inspect our food to make sure it is bacteria-free, we sanitize our water, wash our hands, exercise, eat healthier diets that include a variety of foods, take vitamins, take preventive medicines, vaccinate ourselves, clean our wounds, avoid filth, go to dentists regularly and brush our teeth, thoroughly cook our meats, avoid smoking, and safeguard our work environments. These healthy habits are a major reason that we are so much physiologically healthier and live much longer lives.[25]

In earlier centuries, we did not have the knowledge to apply those health habits described above. Compared to today, we understood little about what harmed us physically. In the realm of psychology, we are just beginning to learn a similar lesson—that psychological health will occur not primarily because we have become adept at curing "mental illness" but because we have learned to practice those skills that enhance our psychological well-being. Our increasing understanding of psychology and neurobiology suggests that the habitual application of healthy practices such as mindfulness and the Three-Step Practice will, we can hope, some time in the future be as unnoteworthy as washing our hands, brushing our teeth, and exercising is today. Similarly, we can imagine that it will be commonplace to notice potentially problematic situational or empathetic emotions that arise, understand what they mean, and act skillfully in light of that knowledge.[26] And we can guess that learning communication skills, relaxation techniques, and how to manage self-beliefs will seem as natural a part of a child's education as personal hygiene is now. We seem poised to move from the disempowering stance wherein we deem most psychological problems to be pathological lightening strikes (random, intense, and when activated, out of our control) to thinking instead of most psychological difficulties as having an understandable etiology that is amenable to our mastery. Good clinical practice suggests that in most instances we are not curing our clients so much as we are teaching them healthy psychological habits that they can continue to apply throughout their lives. If this book proves valuable to people, it will not be primarily because it has furthered psychotherapeutic cures but because it has made a contribution toward the skills and understanding that promote psychological mastery.

CHAPTER 9

Making Unscripted
Emotions Our Allies

In chapter 4, I described a mindfulness exercise that increases awareness of emotions. In that chapter the emphasis was on identifying the particular feelings that constitute an emotion component of a script. I have also noted, however, that mindfulness also increases our attunement to emotional reactions that are natural to our immediate circumstances. These unscripted emotions are a treasure chest of potentially useful information.

Engaging Unscripted Emotions

I have suggested that after you notice an emotion and after you associate the physical sensations of that emotion with an area in your body, you can ascertain the origins of that emotion by asking yourself a few simple questions. The answers to these questions indicate the origin of the emotion you are experiencing—a script; a reaction to your immediate situation or to a situation you are remembering; or an empathetic feeling you are sharing with someone else.

You can check this out for yourself with an experiment. (This exercise is also summarized in the "Identifying the Origins of Emotions" flow chart in chapter 4.) Imagine yourself in a recent situation that evoked some emotion. To the best of your ability, try to identify the area(s) in your body that you associate with that emotion, noting the subtle physical sensations that are caused by it. While repeatedly referencing your physical experience of the emotion, ask yourself the questions below. After each question, pause and reflect upon your answer (reminding yourself to remain attuned to your body).

"Do the emotions seem to fit the situation I'm in (not something additional or different from what I would expect if I looked at the situation objectively)? For example, if someone else were in this situation would I expect him or her to feel what I feel?

"Objectively, would I expect someone else to feel this to the same degree that I feel it?"

If the answer to each of the above questions is "yes," we suggest that you double-check it and make sure. If the answer is still "yes," then your working hypothesis can be that the emotion is a situational emotion.

If the answer to any of the above questions is "no"—if the emotion does not seem to be both proportionate and fitting to the situation, one of the other possibilities is probably coming in to play: either an old script is being activated or you are picking up on another person's emotions, or both.

Check this out further. Imagine the evocative situation again and ask yourself (while staying attuned to the emotion as an experience in your body): "Do these emotions, which I see arising, seem uncomfortably familiar to me? In other words, is this an emotional reaction that seems to stretch back into my personal history?" If the answer is "yes," then you can be quite confident that at least part of what you are experiencing is an old script.

If the emotions seem neither to be accounted for by a script (old and familiar) nor seem to fit the external situation itself, then, consider the possibility that you are feeling what someone else is feeling. Check this out further. Think of the others in the situation that you are imagining. See if what you are feeling seems to fit with what one of the others in that situation would probably be experiencing. If that seems to fit, then the unscripted emotion that you are experiencing is probably an empathetic as compared to a situational emotion.

A good way to practice with these tools is to experiment with assigning a percentage to each category of emotion that you experience in the situation you are imagining. Roughly speaking, what percentage seems to feel old and familiar (scripted); what percentage seems to be the appropriate feelings that the situation engendered in you (situational); and what percentage of emotional feeling seems to be you feeling what another was feeling (empathetic)? After practicing the above exercise a few times you will probably find that you have identified some emotions that are not scripted.

So how does this help you? Nonscripted emotions are rich with helpful insight into our life situations and into the experiences of others.

Using Situational Emotions

We will focus on situational emotions first. Situational emotions inform us about how we are being personally affected in our immediate situation. Once an emotion is mindfully noted and the fact that it is a situational emotion has been verified, then with a few general guidelines you should be able to begin to put those emotions to very good use—or to discard them if they are found useless.

The first thing you want to note is, "What kind of emotion is it that I am having?" There are three general types of emotions that are likely candidates: sadness, anger, and fear. Since our subject matter has to do with human experiences that are problematic, we are leaving out positive emotions. Few people have problems with purely joyful, loving, happy, or excited emotions.

ANGER

In the following pages we will focus more on anger than on any of the other situational emotions because anger is the emotion that most people find most problematic.

We are using the word "anger" to depict a continuum of emotion. We might imagine this continuum to range from "slightly irritated," to "upset," to "quite angry," to "enraged." When we carefully consider the nature of all forms of anger, we discover that anger's function is protective. This makes sense biologically. Evolution did not randomly and gratuitously program our brains to have problematic emotions. Anger has served us as a survival tool. We can say that anger is the naturally arising emotion that energizes us when we perceive ourselves or someone else as getting hurt, and anticipate or remember someone getting hurt. The hurt that we are concerned about may be physical or it may be emotional. If we track any of our angry situational emotions back to their origins we will find that some sort of threat is present, whether it be real or imagined. In the realm of protections that evolution has afforded us, anger represents the "fight" aspect of the flight/fight/freeze protective mechanism.

The protective nature of anger is very helpful to keep in mind because it suggests some very useful questions: "What is the threat that I'm feeling? What is it about this situation that sets off angry alarms? In what ways does it seem that I or someone else is in danger of being harmed?"

Some examples can help us understand this:

Mike is visiting with some acquaintances and they begin debating politics. All of a sudden he is aware of feeling angry. When he asks himself the above questions, he realizes that it seems that if the points he makes in the argument are discredited, then he will be discounted and these people will lose respect for him.

Michelle is driving and another car cuts her off, making her brake hard. She feels a surge of irritation. A close examination of her anger leads her to realize that she feels discounted: how she is affected by having to put on the brakes leaves her with the impression that another driver is marginalizing her importance.

Thomas is conversing with his friends and suddenly he feels irritated. Upon reflection he realizes that the others ignored him when he mentioned to them that he is worried about his health, and so he feels rejected by them.

Tanya is discussing the merits of different career paths with friends and realizes that she is irritated. Upon reflection she notices that her opinions do not seem to be garnering the same respect as those of the men in the group, and so she feels discounted.

Trevor, in talking with a group of friends, hears one of them criticizing another of his friends who is not present. Noting his anger, he sees that he is concerned that this criticism might be hurtful to that friend.

All of these people, by reflecting upon their anger, were able to identify the hurt or potential hurt that underlay their anger. Mike and Michelle felt that they were being discounted. Thomas felt rejected. Tanya also felt discounted, and Trevor was concerned that his friend might be hurt.

We turn now to the next step in working with angry emotions. After we have noticed our irritation and have noted what it purports to protect us from, the next question naturally follows, "Is this anger helpful. In other words, is it useful in defending myself or others from some kind of potential pain?"

In the first two instances described above—Mike equates how much others are going to respect him with how well he does in a political debate, and Michelle feels personally devalued because she has been cut off by another driver—it is likely that those people will see that they really are not at risk. Mike may then decide that a gracious concession of a point can sometimes be more admirable than an overly zealous defense. Michelle might decide that the other driver, who after all does not even know her, is not actually able to diminish her. She decides that she can afford to let go of her anger. Once they realize the function of their anger (to provide protection) and evaluate it as not being useful, people are often surprised at how easy it is to let go of it. In those situations wherein we realize that we do not need protection, our angry emotions are like the child who calms down once she is acknowledged.

The other examples described above suggest a different set of tactics. Thomas, who is feeling hurt because his friends seem disinterested in his health concerns, might recognize his irritation as potentially useful. He might engage that energy and use it to bolster his courage to redirect the conversation toward what is bothering him. He could say something like, "Hey, guys, isn't anyone interested that I'm worried about this problem with my heart?" Or he might confide to one of the friends later that he was surprised that no one seemed interested in his worries.

Tanya, whose opinion is not being valued, also might find her anger to be a useful source of energy for taking care of herself. Perhaps her anger will be applied to assert herself, saying, "I think that my idea about this is a pretty good one. What do you all think?" Or even to say, "I don't think that you guys are truly considering what I'm saying." Similarly Trevor, who was irritated when he heard others criticizing his friend, might could engage his irritation to push himself into the conversation and to state that he does not feel that the others are being fair to his friend.

Sometimes people avoid making an assertive statement like those we have suggested for Thomas, Tanya, and Trevor because they are nervous about how people will react. Upon closer inspection, these concerns tend to fall apart. Using Thomas as our example, it may be that by bringing up his friends' ignoring his comment about his heart, he will learn that they think he worries too much about his health and that he has "cried wolf" too many times. This feedback may be difficult to hear but useful—something that Thomas would not have learned if he had not pursued his irritation. On the other hand, by drawing attention to his friends' lack of acknowledgment he may awaken them to how they have been negligent listeners. Or, we could imagine that Thomas's attempt to engage his friends about their unresponsiveness to his health worries results in the same treatment again, a lack of interest. In this scenario, Thomas may wonder how limited these particular friendships seem to be. He may decide to look to other friendships, especially during those times when he wants more supportive interactions. Each response to his assertiveness, including those that are difficult, leaves Thomas with helpful information that can be readily applied.

Some people might think that these people are overly sensitive; that they feel things too strongly; perhaps these people should be a bit more stoic. This reasoning misses the point. If we lack tools to identify and work with our emotions, then a healthy dose of stoicism makes sense. However, once we have learned to use a few of the tools mentioned here, then our emotions become much less problematic and in fact become useful. As you become more skillful with emotions it makes much less difference whether or not you feel your emotions more strongly or less strongly than others do. Being stoic may be easier for us, but in many situations we find ourselves more enlivened and enriched when we engage and express our emotions.

When we use our anger we will sometimes find ourselves doing things that are a little out of the norm. More often than not, learning to be skillful with anger means being more assertive than we previously knew we could be. Over time, we learn to become more and more skillful in integrating our assertiveness into the more usual flow of conversation, effectively utilizing our anger to protect ourselves or others. Nevertheless, as we become more assertive, we will, at times, behave in ways that are different from what others expect and are used to. The truth is we don't have to be especially deferential and shy about showing up with ourselves—showing up with our emotions and our thoughts. After all, most people in our culture are neither very mindful of their emotions nor adept at applying them. We do not have to adopt their habits as our own.

It is important, however, that when we choose to be assertive we are careful about our motivation. If our motivation is to be heard—to simply acknowledge our experience as equally valuable as anyone else's—then we are on good ground. If, on the other hand, our motivation is to hurt someone or to get into a fight with someone or to prove ourself better than someone else, then the results will probably not be good for anyone, and we might want to take another look at the possibility that an old script is being activated.

BEFRIENDING PERSISTENT ANGER

Most people, at least occasionally, experience angry emotions that do not resolve themselves as easily as is suggested by the examples above. Sometimes, angry feelings persist even after we have applied answers to the question: "In what way do I need to be protected in this situation? What is it that I feel like I need to be protected from?"

More stubborn anger requires a more careful analysis of what the perceived danger is and how to handle it. One way that we suggest doing this—a way that many people have found very helpful—may seem a little odd at first. It involves personifying anger and then negotiating with the personifications.

A personification of anger that many people have found helpful uses the imagery of wrathful allies. This has various advantages for us in examining how to use anger. By personifying your angry emotions, you separate these emotions from yourself—set them out on the table, so to speak, so that you can examine them more closely— and in a better light. Put into neurological terms, personifying emotions allows you to call on the prefrontal cortex and neocortical parts of your brain so that you are better able to reality test and analyze your angry experience.

A carefully wrought personification of anger also reminds us that we are able to act and that we are empowered to protect ourselves. The wrathful allies that we envision as embodying our anger are assertive by nature, not aggressive. However, we do not have to be shy about making them pretty awesome—they have long teeth; instead of fingernails, they have claws; they may hold a sword; they have fierce faces; they are dressed in armor; and so on.[1] They are a metaphor for something that has arisen in our mind, so they are of our same gender. Most important, they function as allies.

These wrathful ones who personify our anger arise to protect us. Even though we may come to recognize their embodiment of our anger as protectively serving us, however, we do not seem to get to decide when our wrathful allies are going to show up. And so, these allies (let's say there are five of them) tend to appear in our experience whenever they feel as if we need to be protected—whether we invite them or not.[2] Nevertheless, we can strategize with these seasoned warriors. Stated differently, we can tame them into becoming powerful and useful, but nonaggressive allies.

To accomplish this, we access a persistent angry feeling and then we imagine these five powerful figures standing behind us, in full battle regalia, ready to protect us. We begin a dialogue with them by suggesting various possible courses of action to them and imagining how they would respond.

Jane provides a good example for us. Jane had had an interaction with her brother and sister-in-law wherein they had been very critical of her—in fact quite rude. She knew that such interactions with them left her doubting herself and just plain feeling bad. Even after she had separated out the old script that they activated and had dealt with it, she was still feeling angry and confused.

She imagined being with her brother and sister-in-law in situations that were sure to arouse the angry emotions. She continued this until she felt the anger in her body. Then she had a conference with her wrathful allies. She imagined them standing behind her, complete with claws, fangs, weapons, anything that made them sufficiently fearless. They were prepared and ready to protect her, no doubt about it!

She began her dialogue by telling her wrathful allies her most recent plan—to end all contact with her brother and his wife. She then sat back and imagined watching her wrathful personifications commenting upon this. (Fritz Perls, that most famous master of personification, might have suggested to Jane that she actually arrange two chairs: one for the wrathful deities and one for herself, moving from one chair to another as she engages different sides of her dialogue. Some people find this very helpful.) The wrathful allies were immediately skeptical that she would actually carry out that plan. Jane quickly realized that they were right. Truth was, she did not really want to reject her brother for the long run. She realized that cutting off contact with her brother was appealing to her not so much as a protection for herself as an attempt to hurt her brother. (Impulses to "get even" are so ubiquitous that it suggests that such actions might have had some evolutionary value—perhaps making our enemies think twice before attacking us again. It should go without saying that in modern society such behaviors are inevitably unproductive and simply create more mutual aggression.)

So she shifted tactics. She suggested to her wrathful allies that she might simply act as if nothing had happened and pretend to be her usual self around her brother, keeping up this act until her hurt and anger subsided. She again felt her allies' skepticism. They pointed out that this would be rather awkward and that she would feel susceptible and defenseless should further attacks occur—not a very good way to take care of herself.

Okay then, she'll call her brother up and give him a piece of her mind, telling him what a jerk he was. She could tell that her allies liked this a little bit better. "Now you're getting on the right track," they seemed to say. But they asked her what would happen after that. She had to acknowledge that either she would end up apologizing for trying to make him feel bad (that again!), or he would simply get defensive and aggressive back, leading to a fight with not much accomplished.

How about if she calls him or has lunch with him and tells him how hurt and angry she is, that attacking her was not a good way to get her to listen, and what was he so upset about anyway? With the help of her allies, Jane sorted out what she might tell her brother that he could hear and take in. The allies were satisfied and they disappeared, but not without acknowledging that they would be there (whether she liked it or not) the next time she needed to be protected. Before Jane talked to her brother, she rehearsed what she wanted to say to him and how she might say it. She imagined her brother's worst possible reaction and how she would handle it. She went over this a number of times before she actually confronted her brother.

People who have practiced these simple steps: noticing when they feel anger; followed by reflecting on what kind of protection, if any, fits their situation; followed by an evaluation of whether that protection is necessary—are often surprised at how quickly they can turn their anger into a useful, nondestructive ally. If you feel stuck with angry emotions, personifying your anger (for example, as wrathful allies) and engaging your personification in a dialogue is often helpful.

SADNESS

Another category of situational emotion is sadness. Sadness informs us that something that we have valued is either gone or threatening to go. What do we do in such a case? Generally two things are suggested.

The first is to allow ourselves to grieve. Small grief may last only a moment, while large grief may last years. Regardless of the size, skillful grieving always has as its basis the cultivation of a kindly, warm concern for oneself. By accepting our grief we can eventually come to feel moments of celebration for our good fortune in being able to have had something that was precious enough that we miss it so. Many good and helpful books have been written about grieving, and so here we will only a few brief comments.[3]

Grieving is, of course, a natural reaction to loss. As we go through the stages of grieving it is helpful to remind ourselves that our sadness is occurring in our bodies and that our bodies can be trusted. Sometimes people are frightened of the strength of their grief and fear that it will overcome them. It is true that in the throes of sadness there can arise moments when we experience ourselves as disappearing—buried and subsumed by the sadness. But this is no reason for panic. We always come back. When waves of sadness arise and engulf us, after sobbing or otherwise riding the sadness out, we resurface. Although we might disappear for a moment, we will reconstitute ourselves after the sadness has had its time. Usually, after crying we find moments of calm.

Everyone grieves differently and different losses represent different courses of grief. Grief becomes especially problematic—we are most likely to get stuck in it—when it activates old scripts. For example, in the midst of our grieving we might become stuck in self-criticism; we might become habitually embittered toward others, or we might maintain an image of the world as an uncaring hostile place. It is common for grief to activate old scripts like these. As the grief subsides, most people move on, and the scripts return to a more latent status. Sometimes however, loss activates scripts that stubbornly stick around and cause the grief to be complicated. Grief that becomes stuck represents a call to use the steps described in this book to identify and master an amygdala script or to get outside help if that does not work.

I mentioned that two things are suggested for grief. The second is that grief can be connecting. Grief is a close cousin to love. In fact, it probably cannot exist without love. After all, we have to find something lovable in order to miss it! Perhaps this is why sharing sadness is usually very connecting. Whenever we notice that we are feeling sad, we should consider the possibility of sharing our emotion with someone else. More often than not, this ends up psychologically beneficial to both.

Another situational feeling of sadness is bittersweet sadness. This emotion connotes moments of great appreciation for that which we value—a person, a group of people, a place, maybe even just waking up to a beautiful morning. Strongly felt appreciation can have a sad quality because in the background there is the reality of impermanence. Our appreciation is heightened by the certain knowledge that each moment is also a fleeting one. Hence, when we feel wonderfully touched by something it is often a rich mixture of happiness and sadness. The incredibly bittersweet quality of saying goodbye to a loved one is an example of this.

Other examples are common. A father returns home from work and his two little boys run up to greet him, and he simultaneously hugs both of them. This happens almost every day, and each time it is a wonderful experience for the father and his sons. Wonderful and precious because the father knows, and the boys intuit, that it will not always be so. Some day the boys will grow up and some day the father will die. If this were not the case, if the father and the sons were certain that they would be able to greet each other this way forever, then there would be nothing precious about it. Why bother to embrace our children today if we know for sure that we will be able to do it every day forever? Knowing that what presently enriches our life will at some time disappear makes things very precious. Bittersweet sadness or poignancy is born out of our certainty that nothing is permanent, and so, when seen clearly, this type of sadness is recognized as a celebration of what is precious to us. It is perhaps the richest kind of sadness because it acknowledges that which we value most, while simultaneously accepting, maybe even without complaint, that all moments and things will pass.

FEAR

The third general type of situational emotion that we mentioned is fear. After we have separated out the part of our fear that is situational (not an old script and not the result of an unconscious empathy with someone else), the first step toward the skillful engagement of fear is to consciously and deliberately ask ourselves the simple question: "Am I really in danger?" We emphasize "consciously and deliberately" because a passing nod to the fact that you are not in real danger does not sufficiently alert the neocortex to assert itself. A very deliberate statement to yourself, even to the point of answering the question out loud, is more likely to engage the neocortex and its reality testing functioning: "Right now, right here, I am in no danger and so need not be afraid." Such an explicit verbal statement is much more effective in calling upon the neocortex than is a more vague, implicitly acknowledged awareness that there is no real danger. This statement—a reminder of one's immediate safety—may need to be repeated a number of times. With fear it is also important to be especially conscientious in carefully separating out any old script that might be part of it and dealing with that script first.

Unlike anger and sadness, in the case of fear there are few situations in modern life, at least in peaceful countries, where fear actually fits; in most places most of the time we are not in any real, imminent danger. That statement may seem surprising at first. Much of the media thrives on portraying the world as a dangerous place—making prominent stories about random acts of violence, the most horrific of accidents, and a variety of other unforeseeable tragedies. Media moguls know that although these things may not be technically newsworthy they draw readers and viewers. Governments and politicians also habitually exaggerate danger in order to get us to support their views and policies. We are bombarded from many directions by an exaggerated vision of the dangers of the world, but acknowledging this is not meant to deny that, at times, it is wise to be especially vigilant—while driving a car, for example. Being alert and vigilant is different, however, from feeling fearful. Even if we are lost in the mountains or face criminal activity in urban areas, it is better to reassure ourselves by noting that at this moment we are not in immediate danger; a calm mind is much more likely to make good decisions about how to proceed. If, in our vigilance, we do see an imminent danger—a car swerving toward us, a suspicious-looking person changing course toward us on a darkened street—we can count on our bodies to provide us with the adrenaline that we need in that moment to act quickly.

If the answer to the question "Is there really something in my present situation that I need to fear?" is "no," then, assuming that an old script is not still getting in the way, it should be fairly easy to reassure yourself. In reassuring yourself, you are activating the neocortex part of your brain to moderate an amygdala-based reaction. Remember that the language of the cortex is just that—actual language. Therefore, in reassuring yourself it is often helpful to use actual words, "I am really not in danger here." Practicing a relaxation training method such as is described in appendix A is also very helpful.

If, however, the answer to the question "Is there really something to be afraid of here?" is "yes," then (unless the danger is imminently obvious, in which case usually the best next thing to do is to get the hell away from wherever you are!) then it is helpful to ask the next question: "How big a loss am I facing?" For example, if my friend rejects me: "Will it be forever? Will I find other friends?" If I know that my coworker is going to be angry at me: "What is the worst thing that could happen? What would I do then?" Those two questions can be very helpful in many contexts when we are faced with the situational emotion of fear—"What is the worst realistic thing that could happen?" and if it did happen, "What would I do then?" and "After I've done that, what is the worst realistic thing that could happen?" and so on. Applying a series of these questions leads to the realization that, in the vast majority of situations, if the fears that we have came to pass we would still find the situation workable. The erroneous message that fears often give us is that we should be alarmed and very worried, as if our very survival is at stake. Closer examination of the actual contents of our fears can be very reassuring. After answering these questions and thereby throwing a more realistic light upon what is occurring, it is very helpful to remind yourself repeatedly of the bottom line—you will survive and you will find a way.

Arthur provides a good example of this. Arthur is an alcoholic. Before becoming sober, Arthur had abandoned his wife and son, choosing alcohol as his primary love. He had been sober for five years when his twenty-eight-year-old son agreed to meet with him. Arthur was eager to meet his son and also terribly afraid. He had done a great deal of work regarding his guilt and making amends, but still he understandably feared the worst in terms of his son's reaction to him. By asking himself, "What is the worst that could happen?" he realized that as painful as a rejection by his son might be, it would still be survivable. His life would still go on and he would not lose some of the things that he had built up during his sobriety. Also, a rejection at this time would not necessarily mean a rejection forever.

Arthur began practicing for meeting with his son. To start a practice session, Arthur would relax his body. He would then envision himself being with his son, imagining ways in which it might not go well and various possible responses that might help him, reminding himself that no matter what happened his life could go on in a satisfactory manner.

A side benefit was that as a result of his rehearsals for their meetings, Arthur discovered that there were a few key things that he wanted to communicate to his son: "This is the most frightening thing that I think I have ever done. I know you will have some negative emotions toward me, probably strong emotions. . . ." Numerous times he practiced imagining the worst possible outcome, and each time he reminded himself that he would survive it. He also practiced approaching the meeting with an attitude of confidence, an expectation that it would go well. It was made easier to carry that vision because he knew that he would be all right even if it went horribly. He went to the meeting with an attitude of confidence, an image of how it could go well, and a practiced certainty that if it went badly he would be all right.

When the first meeting actually occurred the result was mixed. Arthur's son was understandably angry and was mostly reserved, responding minimally to Arthur's endeavors. However, Arthur felt good about the fact that his many rehearsals of this meeting allowed him to separate out, on the spot, the old scripts that were being activated for him and to respond to his son in an understanding and empathetic manner. At last report, Arthur had met with his son three more times. Each time had been hard, and each time Arthur left feeling that he had connected with his son, though not much of a sense that his son was connecting with him. He nevertheless felt that with perseverance there might come a day when they could interact in the way that Arthur yearned for. "If not," Arthur stated, "at least I know that I can love my son, give some to him, and I am very proud of him. I'm increasingly hopeful about us because I know that my son has the courage to keep showing up even though this is also very difficult for him."

Summary of How to Work with Situational Emotions

Working with situational emotions entails, first, noticing the emotions as sensations in your body, then identifying the category of the emotion that is activated (angry, sad, or fearful). This is followed by reminding yourself what the function of that emotion is (anger, as the perceived need for protection; sadness, as the acceptance of loss; fear, as a perceived danger). Finally, we must apply the various questions and strategies mentioned above so that we can be most skillful in relating to our immediate situation.

This may sound a bit cumbersome and not very spontaneous. After a bit of practice, however, these guidelines become automatic and actually create the potential for more spontaneity. There is a musical analogy. Each type of emotion could be thought of as a different musical key (like B flat, E major, and so on). With problematic emotions we only have three keys: the key of anger, the key of sadness, and the key of fear. When we practice identifying emotions and naming them, it is like we are practicing our scales on a musical instrument, each scale representing a different key and each key representing a somewhat different flavor of expression. Musicians practice playing the scales of each key, making it easy for them to recognize a particular key, and to improvise in the context of that key. Practicing with emotions awhile similarly promotes quick recognition of an emotion and what it means. When that emotion arises in our daily lives, our familiarity with it allows us to improvise creatively within the context of the meaning of that emotion.

Empathetic Emotions

Along with alerting us to scripts and attuning us to how we are affected by a present situation, mindful attention to our emotions can also inform us about the emotions of others. Arthur provides an example of this in his meetings with his son. One of the things that helped Arthur during his interactions with his son was his seemingly unusual ability to tune into and appreciate his son's emotions. This did not occur for Arthur because he was born with an unusually acute ability to empathize with others; it was the result of Arthur's focused practice on the nuts and bolts of empathy.

Researchers have shown that when one person is asked to empathize with the emotions of another, they both show similar activity in the amygdala regions of the brain. This, and other research, has actually spawned a new area of empirical research—"emotional intelligence."[4] While the concept of empathy has long been part of the clinical lexicon, only recently has science developed better ways to measure it directly, making empathy the subject of a considerable amount of research.[5] A likely mechanism for empathy has been suggested by researchers who are studying "neural mirroring." Neural mirroring occurs when, by observing another person's emotions or behaviors, "mirror neurons" become active in the same areas of the observer's brain that that are active in the brain of the person being observed. Stated differently, when we are around someone—especially if they are having a fairly strong emotional experience—mirror neurons become engaged in the same regions of our brain that would be active if we ourselves were having the experience.[6]

Because emotional empathy represents one of the functions of the amygdala, it is not surprising that the process through which empathy occurs, like most of the processes associated with the amygdala, is primarily unconscious. For example, people can be shown to be quite good at empathizing yet they often are not consciously aware of exactly how they are doing it or even that they are doing it. It seems that this happens commonly.[7] One of the characteristics that makes life so confusing and difficult for people with autism is that they lack the ability to utilize mirror neurons to empathize emotionally with others.[8]

As we began teaching people to tune into their bodies as a means of identifying the emotion component of scripts, we discovered that most people could quickly learn to separate out what was an old scripted emotion from what were present situational emotions. We had expected that. What was more surprising was how often people discovered that the emotions in their body mimicked the emotions of those around them. We began to encourage people to try separating the emotions that were their own (either situational emotions or emotions spawned by an old script being activated) from the emotions of those around them. We found that with only a little practice, most people could also easily make that distinction.

This probably should not have surprised us. It is commonly accepted that if people are asked how their spouse, child, friend, or even acquaintance felt about something, they can very often report a fairly accurate sense of the other person's state of mind (such as, she didn't like that, it seemed to irritate him, or it made her nervous). It is noteworthy that the above example depicts the reporting of another's emotional state some time *after* the situation with that other person occurred. Although we may be neither consciously trying to be empathetic nor even consciously aware that empathy is occurring, we nevertheless seem to be recording the emotional states of others.

To experiment with how we can feel the emotions of others, think of a conversation that you have had recently, a conversation that seemed to be somewhat emotional. Think of the other people who were involved in the conversation. Focus on one of them. As you imagine that person in that situation, maintain an awareness of your body. Actively focus on the image of that person in that situation while passively maintaining an awareness of the sensations in your body. Now, ask yourself, "What does it seem like that person was feeling?" Practice this for a minute or two.

After practicing this exercise a few times, people generally report that they are able to notice emotions that they would otherwise have either ignored or attributed to themselves. As they continue to apply the above approach (namely, staying attuned to their subtle bodily sensations while asking themselves, "What is this person's emotion?") they find that what is going on in a given situation becomes much clearer and results in greater empathy and compassion for those with whom they interacted.

FIVE CAUTIONARY NOTES REGARDING EMPATHY

1 If your empathetic experience of someone else does not result in a compassionate and accepting understanding of that person (even though you may still disapprove of how they were behaving), then you have missed something.

2 It perhaps goes without saying that what you experience empathetically with someone represents only a working hypothesis. Be conservative about solidifying another's experience into a certainty. Check out what seem like empathetic emotions. Do the emotions that you seem to be experiencing via the other person fit with what you know about that other person? Do they fit with the behavior that you observed?

3 Remember that we are talking about emotional insight, insight that comes from the amygdala. Amygdala-based emotional empathy is often very accurate, but it is the neocortex that puts it into words. The part of the neocortex that concerns itself with language and concepts has little to do with empathy. As a result, the words that come to your mind may not fit the words that the person sitting across from you might use to describe his or her experience. This could be true even if the person is having the exact emotion that you are empathetically experiencing.

4 Furthermore, the context for the emotion may also be different. For example, the person across from you may in fact be having the angry, fearful emotion that you have noted in your body, but what you call anger and fear your friend may simply think of as confusion. Also, you may have thought that your friend's anger is somehow associated with you, but find that you have to reevaluate this when it turns out that he or she is upset with a boss at work. Your body may inform you wonderfully about the quality and intensity of what your friend's emotion is—freeing you up to be especially present and tuned in— but it tells you little about the language and arena in which that emotion exists.

5 Do not feel that simply because you have some insight regarding someone, you should then share it with them . . . or otherwise use it to further enlighten them. Your desire to do this might well represent a conceit and your insight, even if it is accurate, may not be welcomed. No good comes from telling anyone anything that they are not prepared to hear. And if, after empathizing with someone, they indicate to you that their experience was not what you perceived, honor that. It almost always marks the height of arrogance ("therapeutic aggression" if you are a therapist) to disagree with someone about their experience. This applies even if you remain quite certain of what you have experienced via them.

———

This concludes our brief discussion about how unscripted emotions— those emotions that are circumstantial and those that are empathetic— may be identified and made useful. The subject could easily be the basis for another book, but we hope that the reader will at least have had his or her curiosity piqued. Of course, we encourage people to further explore this subject, both experientially, through practicing the exercises mentioned above, and also by exploring the rich body of literature that exists regarding mindfulness and empathy.

———

We have come to the end of this exploration of amygdala script theory and the Three-Step Practice. It is my very good fortune to have the pleasure of sharing this with you. Many people have contributed to what you have just read—some directly, some more indirectly through their fearless embrace and mastery of the human mind. I share these ideas with you, hopeful that you will find your life enriched by the practices and insights described above. I would also like to encourage readers to consider this book—grounded in the scientific and empirical traditions of our modern world—to be an invitation to further exploration of those other, more subtle traditions, paths of mindfulness that have been refined for thousands of years.

RELAXATION EXERCISE

Below is a modified version of a stress-reduction technique that was first taught to me by Johann Stoyva in the 1980s. Johann, a meticulous researcher, could well be considered the father of modern biofeedback. His studies in the 1970s and 1980s had shown that relaxation had three facets: the autonomic nervous system, muscles, and cognitions or thoughts. Armed with this knowledge, he borrowed from a variety of existing relaxation tools to develop this simple exercise that addresses each of these arenas. He then refined and tested the result using biofeedback techniques.

THE EXERCISE

☑ Begin by adopting a good posture—aligning all of your vertebrae so that your backbone becomes an effortless support for the rest of your body.

☑ Imagine a place that is very relaxing—perhaps a mountain meadow, a beach, your bedroom, floating on a cloud, a field—anywhere that you deem especially peaceful. Imagine that you are very safe and you will not be bothered there. It is as if the place itself welcomes you and will take care of you.

☑ The following is a breathing technique that is common to yoga, singing, playing a musical wind instrument, and sleep. Begin by breathing in and let the incoming air push your stomach out. The in-breath should require virtually no activity in your upper chest: let your stomach muscles do all of the work required to inhale. Breathe only a little more deeply than usual. Hold each breath for ten seconds or so, then relax your stomach muscles. Continue in this fashion, letting the air flow in and out at an easy pace, gently using only your stomach muscles. With a little practice you will likely find a rhythm that seems comfortable—even familiar. This is because we naturally resort to this type of breathing as we sleep. The technique has various names, including "abdominal breathing."

☑ With each exhalation think or say the word "relax." Imagine that all of the tension in your body is released with your breath, flowing down through your body into the ground.

☑ To relax your muscles begin with your feet. Bring your attention to the muscles in your feet. Notice any tension that may exist in those muscles. Envision your feet muscles relaxing and any tension that might have been held there flowing down into the ground. Move your awareness to the muscles in your calves. Focus upon each calf, noticing any bit of tightness that might exist there, and let that tension move down through your feet into the ground, as your calf muscles become more and more relaxed. Proceed in this way to address almost every muscle group in your body: lower thighs, quads, buttocks, lower back, back, chest, biceps, triceps, forearms, hands, upper back, shoulders, neck, jaw, mouth area, forehead area, and scalp. Occasionally remind yourself to use the abdominal breathing. Often there is a muscle group that we tend to habitually tense. Don't become discouraged if particular groups of muscles won't relax, only partially relax, or quickly resume a state of tension after relaxing. Sometimes it is helpful to tense and release muscles that are especially stubborn, but don't be too concerned—just do the best you can with these groups and move on. Eventually these muscles will learn to become more at ease.

☑ After having moved through your body in this way, return to the peaceful place that you envisioned above and rest there. Continue the abdominal breathing and with every breath out continue to say or think the word, "Relax." Notice and enjoy the feeling of relaxation. You can use the memory of this feeling to help relax yourself throughout the day.

Try to set aside time to practice this exercise every day. Relaxing oneself is a skill—you get better and better at it the more you practice it. You will find after awhile that you can go through the whole exercise in fifteen minutes or so, but the longer you spend the more effective it is.

People spend thousands of dollars going to exotic places, partly in hopes that their stress will be reduced. Regularly doing the steps above will be more enduringly effective than the vast majority of vacations. Of all of the ways that we might use our time, relaxing is one of the more beneficial and enjoyable.

Throughout the day, if you feel tense or stressed, try the abdominal breathing technique described above. In fact, if you breathe with your stomach at all times you will find that you maintain a more relaxed state, more present and skillfully responsive to whatever arises. Don't stress yourself trying to accomplish this, though!

It is also a good practice to do a body scan periodically. This entails quickly scanning through the muscles of your body and, upon noticing any area where there is a bit of tension, imagining that the tension releases and flowing down through your body into the ground.

_____ NEUROBIOLOGICAL RESEARCH NOTES

For the sake of brevity, I have limited these notes to the neurobiological research that supports those concepts most central to amygdala script theory.

The Amygdala's Role in Psychological Problems

The amygdala is generally recognized as having an especially important role in a variety of psychological problems.[1] This region of the brain is often center stage in neurobiological research on anxiety-related disorders,[2] depressive disorders,[3] borderline personality disorder,[4] and even addictions.[5] Although the amygdaloid areas of the brain always act in concert with many other areas of the brain, this region maintains a conspicuous role in the creation and activation of most psychological problems. This suggests "amygdala" as the most prominent candidate for a term to symbolize those subcortical processes which underlie such difficulties, hence my frequent use of terms such as "amygdala-mediated memories," "amygdala-mediated processes," and most important, "amygdala scripts."

The latter term arises because I have posited that fundamental to the way in which the amygdala mediates the activation of problematic psychological emotions, perceptions, and behaviors is through a "script." A script is depicted as a learned response, unconsciously encoded and activated, and distillable into three components: an emotion component, an image component, and a belief component. Below is a review of some of the literature that supports each of these characteristics.

The Unconscious Nature of an Activated Amygdala Script

We turn first to our formulation of an amygdala script as learned and activated with little conscious awareness. Many researchers have underlined the unconscious nature of an activated, previously learned, amygdala-mediated emotional response.[6] Recently, researchers have preferred the phrase "automatic emotional processing" rather than "unconscious emotional processing"—with no change in meaning.[7] The studies cited above make it clear that the neurological sequences that initiate and enact the retrieval of an emotion memory occur "automatically," heedless of the purposeful direction or conscious awareness that is typified in higher cognitive levels of processing. (We shall see that this does not preclude the possibility that neocortical regions of the brain can be marshaled to support or, as in the case of the Three-Step Practice, to consciously modify a subcortical emotion response.)

Learned, Long-Term Amygdala-Mediated Emotion Responses

We have emphasized that those amygdala-mediated responses central to an amygdala script are learned and take the form of a long-term emotion memory. Joseph LeDoux was the leader in demonstrating the amygdala's role in the creation of emotion memories. His work has subsequently been verified, underscoring the amygdala's crucial role in the encoding, consolidation, and retrieval of emotion memories.[8] However, the amygdala may take a role secondary to hippocampal regions of the brain in the retention of such memories.[9]

Concerning the long-term nature of these memories, Florin Dolcos, Kevin LaBarr, and Roberto Cabeza demonstrated experimentally the continued existence of durable, amygdala-mediated memories for a period of a year.[10] Natalie Denburg, Tony Buchanan, and colleagues describe research that further supports the robust nature of this type of memory, showing that emotion-laden memories remain intact over a lifetime, whereas declarative memories tend to deteriorate.[11] Other researchers have suggested the same,[12] and a review of research addressing this question has inspired at least one normally staid researcher, to quote the rather dramatic metaphor of William James: "the 'scar upon the cerebral tissues' that characterizes emotion's impact on human . . . memory."[13]

We discussed in chapter 3 the possibility that an emotion memory might be learned vicariously, through empathy with others. Elizabeth Phelps offers a 2006 summary of a variety of research showing the amygdala's role in empathy and attunement to the emotions of others.[14] So it is not surprising that amygdala-mediated responses can be conditioned when one experiences the strong emotion of a primary person in his or her life. Arne Öhman and Susan Mineka provide a review of research that supports this regarding amygdala-mediated fear reactions.[15]

The Emotion Component

Another noncontroversial aspect of the amygdala region's functioning is the activation of emotions.[16] The mediation of psychological problems and the mediation of emotions in the same area of the brain does not necessarily imply, however, that these two aspects of amygdala activity are related. Although emotions have long been assumed to have a causal relationship with psychological problems, the neurobiological justification for such an assumption is another matter. In particular, are we justified in our emphasis on the emotion component of an amygdala script as being an important contributor to psychological problems? Writing on a related topic, Rudi De Raedt cites a large body of research that convincingly ties depressive disorders and anxiety disorders to dysfunctional regulation of emotional material.[17] This correlates with uninhibited amygdala activity. Mclissa Green and Gin Malhi review a somewhat different body of research and come to a similar conclusion: "The capacity to effectively implement emotion regulation strategies is essential for psychological health."[18] Not surprisingly, the research they cite centralizes the tasks that the amygdala regions of the brain play in the generation of problematic emotional content. Richard Davidson, Darin Jackson, and Ned Kalin, in another review of research pertaining to the neurobiology that underlies psychological problems, conclude that emotion "is a key component, if not the major ingredient, for many of the fundamental dimensions of personality and vulnerability factors that govern risk for psychopathology."[19]

Our concept of an emotion component also suggests that once activated, the emotional response is not easily terminated or modulated. In a review of relevant neurological research, Öhman and Mineka, "Fears, Phobias, and Preparedness: Toward an Evolved Module of Fear and Fear Learning," state that once an amygdala-mediated fear response is activated, it "tends to run its course with few possibilities for other processes to interfere with or stop it."[20] Phelps, in a review of the amygdala's role in impacting cognitions, concludes similarly: "When attention is captured by emotion, there is impaired processing of nonemotional aspects of the stimulus or event."[21] It requires unusual interventions, such as the Three-Step Practice, to moderate the effects of an activated amygdala script.

Mastering the Emotion Component of a Script: Step 1 of the Three-Step Practice

It is now well established that the prefrontal cortex, in concert with other related areas of the brain, can be activated to moderate the activity of the amygdaloid complex. Reduction in amygdala activity correlates with reduction in the intensity of emotional reactions.[22] This also reduces the tendency for emotions to narrow the focus of attention and distort present realities.[23] One way in which the prefrontal cortex can be activated to moderate affective responses is through mindfulness practice, including mindfulness practices specific to the identification and naming of emotional states. Studies such as those achieved by Creswell and colleagues and that of Lieberman and colleagues provide neurobiological support for the conclusions of Brown and Ryan, who review an extensive body of literature that demonstrates the salutary effects of mindfulness practices, including those similar to step 1, wherein the focus of awareness is upon internal emotional states.[24]

The form of mindfulness that is learned in step 1 entails repeatedly identifying emotional states to condition a complex of prefrontal and related areas of the brain to respond to emotional stimuli in a manner that promotes emotional modulation. One of the key players implicated in this process is the insula region of the brain.[25] Interestingly, there is evidence that over time, mindfulness practices serve to increase the gray matter in the anterior insula and other related areas of the brain, promising a variety of long-term benefits, including skillfulness in affect modulation.[26]

Neurobiological Basis for the Image Component of a Script

It has become well accepted that the amygdala plays a central role in the conditioned learning of emotional responses.[27] Conditioned responses are activated by particular stimuli and not activated by others, hence learning a conditioned response includes the encoding of some aspect of a context that serves to activate that response, namely, learning the cues that activate a "context-dependent emotional response."[28] In order for a conditioned emotional response to be useful, it must not only be context-dependent but it must also be generalizable to other similar but different contexts. If you are bitten by a dog you will probably incur a conditioned response that causes you to be wary of dogs. To be useful, this learning must generalize from your experience with a particular dog to include a wariness of certain types of dog behaviors in a variety of contexts and with a variety of dogs. Without this sort of generalization, you would only have learned to be wary of one dog and its particular behavior, engendering very little aid in safeguarding you from future dog attacks. The amygdala solves this problem by mediating the encoding of a general outline, but not the details of the circumstances surrounding the original learning situation. Therefore, when future similar situations activate our conditioned response, our emotional reaction occurs in concert with additional information—a general outline or image of how the present situation might be harmful to us. This outline includes information derived from the original learning situation but lacks those details specific to the original learning situation.[29] This undetailed information that allows a conditioned response to generalize has been referred to as an "implicit memory," a "gist," and memory "intrusions" of previously learned imagery and emotions.[30] We have chosen to use the term "image" because, with only a little explanation, most clients readily understand the applicability of this everyday word.

Neuropsychologists Jacobs and Nadel offer a model for describing the process whereby an amygdala script encodes an image that is nonspecific and devoid of detail: "Because the structure critical to forming and storing the contextual frame (the hippocampus) is disrupted by the hormonal cascade initiated by stress, this aspect of memory [autobiographical detail] is lost. Because the structure critical to forming emotion memories (the amygdala) is enhanced by this same stress-induced hormonal cascade, emotional hypermnesia results. Putting these two together would yield strong emotion memories divorced from their appropriate frame of reference" (brackets mine).[31] Returning to our example of being a dog-bite victim, we can easily imagine that when you are now approached by a dog, your body will react, readying itself for fight, flight, or freeze behaviors. This constitutes the emotion component of your script. Simultaneously, you can be expected to experience a general, undetailed image of how the present dog might harm you. It is this lack of detail that allows you to apply what you previously learned when bitten by a dog to your present circumstances.

**Mastering the Image Component of a Script:
Step 2 of the Three-Step Practice**

In Step 2, historical associations are made and then conditioned to an amygdala-mediated response to produce moderation of that response.

Rudi De Raedt and others have highlighted how hippocampus activity is associated with the moderation of emotional and cognitive distortions correlated with heightened amygdala activity.[32] The hippocampus is best known in its activity associated with the recording of episodic or declarative memories—those memories that include conscious recall of historical occurrences. Although the hippocampus may also play a role in the unconscious recall of an emotion memory, its role is certain and central to the conscious recall of memories. The hippocampus also mediates the conscious awareness of context so, for example, when subjects are asked to recall emotional autobiographical memories, the hippocampus and the amygdala are both activated, as is the medial prefrontal cortex.[33] During the conscious recall of memories, hippocampal activation has been shown to moderate an emotional, amygdala-mediated response.[34] Phelps notes that "The hippocampal complex, by forming episodic representations of the emotional significance and interpretation of events, can influence the amygdala response when emotional stimuli are encountered."[35] Furthermore, research suggests that when an emotional event is recalled, that memory has been shown to be especially malleable, subject to modifications that could, among other things, moderate its problematic aspects. As Buchanan states, "The malleability of the memory trace following retrieval, whether induced by CBT [cognitive behavioral therapy] or pharmacological means, may thus allow for the amelioration of traumatic memories in individuals afflicted with psychopathology."[36] By utilizing the hippocampal regions of the brain to tie conscious recall of emotion-laden memories with a historical context, step 2 would seem to induce this heightened malleablility, and capitalize upon it as an opportunity to moderate amygdala-mediated emotionality.

It seems justified to conclude that, in the second step, when clients practice associating an activated amygdala script with a historical memory, they are conditioning salutary hippocampal and prefrontal cortex responses that ameliorate the intensity of an activated amygdala script. Or, in simpler terms, the image component of an activated script becomes consciously tied to its historical origins, thereby reducing its emotionality and taming its dysfunctional tendencies.

It is interesting from a scientific standpoint to note that there is evidence that the behavioral conditioning such as is practiced in the activation of the hippocampus in step 2 might create psychological improvement, not just through the long-term potentiation (LTP) of existing neural networks, but also through neurogenesis, the actual creation of additional neurons (gray matter) in the brain. In a review of relevant research, Davidson and colleagues note that "Collectively, these findings highlight the plasticity of certain regions of the brain that persists into adulthood and raise the possibility that interventions, even those occurring during adulthood, not only can have effects on neuronal function but also can literally influence neurogenesis. The fact that the focus of this work is in the hippocampus indicates that a major substrate of context-dependent emotional responding is a key target for these experientially induced changes."[37] De Raedt cites additional research that supports this possibility.[38]

Mastering the Belief Component of a Script: Step 3 of the Three-Step Practice

Step 3 emphasizes a cognitive reappraisal of an activated script and conditions that appraisal to an amygdala-mediated response. It accomplishes this by conditioning new meaning to old emotion-laden memory. In step 3, clients reframe the emotion component of a script into a context that disavows the present relevance of its emotional content. Green and Malhi note how this form of reappraisal is especially effective in allaying the intensity and also the behavioral problems associated with emotions. They also describe some of the neurobiological underpinnings of the moderation effect that occurs when an emotional reaction is (consciously) reappraised as not relevant to one's present reality. They cite a body of research that demonstrates how this type of reappraisal enlists the medial prefrontal cortex—an area that is involved in monitoring and evaluating certain types of internal information.[39] Phelps, in "Emotion and Cognition," elaborates on this, citing additional research to suggest "that conscious, emotion regulation strategies, which depend on lateral PFC regions known to be important for executive processes and working memory, may act to diminish negative emotional responses by virtue of their influence on medial prefrontal cortex regions that have been shown to inhibit the amygdala during extinction."[40]

Additional Note Regarding the Neurobiology That Underlies Conscious Awareness of Emotional States

I have suggested that focusing on subtle body manifestations of an emotion aids in being able to differentiate between types of emotions and their origins. More specifically, I have noted that the autonomic nervous system releases hormones to create an emotional reaction that primarily manifests in the body. I also suggest that emotion-caused changes in our internal body are detected by regions of the brain that monitor internal physiological states, thereby promoting our awareness of emotions. The data supporting this are strong.[41] What is less clear is whether the brain utilizes subtle autonomic nervous system differences in the peripheral body to differentiate between emotions and their origins or if, instead, the brain utilizes an as yet unidentified process that occurs solely in the central nervous system to make the more subtle distinctions between a particular emotion and its origins vis-à-vis other emotions.[42] Either way, there is considerable evidence to support the therapeutic value of focusing upon subtle physical manifestations of an emotion.[43] It makes little clinical difference whether this attunement is directly responsible for the discernment of subtle differences in emotion as they manifest in our bodies or if such practices, by creating heightened conscious awareness of emotional states, serve to catalyze the brain's ability to make those distinctions more centrally.

_____AMYGDALA SCRIPTS AND CHILD CARE

Amygdala script theory offers pointers to those of us who are responsible for children. For example, we do well if we support and empathize with a little one who has been upset by something and wants to tell the story again and again. Repeatedly describing a traumatizing situation actually serves to reduce the intensity of an amygdala script. We can enhance the healing effect of this by commenting that what is being described happened in the past and is now a memory—but not in a voice that discounts or minimizes the importance that a child attaches to the story. While the child is telling the story—while the script is active and amenable to modification—we are validating the child's feelings, while at the same time reminding him that what he is experiencing is a memory of an emotion that occurred in the past. In psychological terms, we are calling upon the neocortical regions of that child's brain and conditioning the experience to be associated with a conscious memory of a historical event. The goal is to increase the likelihood that if a script has been created and becomes activated, it will be less intense and easily mastered, insofar as the ensuing emotions will be more readily recognized as belonging to a memory that is not relevant to present circumstances.

Humans, especially in their interactions with children, seem to have a natural desire to comfort others who have been traumatized or repeatedly hurt. We instinctively do this in a manner that seems to address the three components of an amygdala script. For example, imagine a child on a playground who has just been hurt by a bully and a teacher who has come to the child's aid, or imagine a child who has been traumatically rejected by her girlfriends and a parent is comforting her.

"Are you okay? I'm sorry that you got hurt so." _(The caretaker puts her arm around the child to provide comfort and acknowledging that being bullied or rejected makes us feel bad.)_

"Those kids were mean to you. I remember how awful I felt when kids were mean to me. Kids are just mean sometimes, but that doesn't mean that there is anything wrong with you. It happens to everyone. It's not your fault."

These spontaneous responses to a child's upset can diffuse a potentially harmful situation. Such endeavors represent a rendition of the three steps: (1) identifying a painful feeling (mindfulness); (2) noting that the feeling was caused by the particular circumstances that occurred when the script was instituted (historical insight); and (3) changing the dysfunctional beliefs about oneself or about the world that were construed from the original situation (cognitive change). A close look at the above playground incident provides guidelines for how to handle such situations by highlighting the three steps:

1   *"I'm so sorry that you got hurt."* This is an acknowledgment of the painful feelings that the child is having. Creating this conscious awareness gently directs a child's attention toward an acknowledgment of his internal experience and reassures him that he does not have to be afraid of these negative emotions. This can be seen as a form of the mindfulness practice of step one of the Three-Step Practice.

2   *"Those kids were mean to you. I remember how awful I felt when kids were mean to me."* By emphasizing what caused the child to feel so bad, the caretaker is offering a context for what has happened—insight into the origins of the child's hurt. This serves the same purpose as the historical insight afforded by the second step of the Three-Step Practice.

3   *"Kids are just mean sometimes, but that doesn't mean that there is anything wrong with you. It happens to everyone. It's not your fault."* The caretaker is providing a useful belief ("there is nothing wrong with you") to allay any dysfunctional beliefs that could become solidified in a script—similar to the cognitive change of the third step in the Three-Step Practice.

Also, when the caretaker mentions in the previous sentence that something similar has happened to her, she is suggesting that such incidents are survivable, potentially allaying a belief that could lead to a heightened sensitivity regarding future rejections. The caretaker's warmth provides reinforcement for taking on a more functional belief about oneself.

Here is a summary of the three steps that we can offer a child who has been harmed: acknowledgment of the child's hurt, acknowledgment of what caused the hurt to them, and comments designed to allay the child's future fears or other dysfunctional beliefs that might be in danger of forming. We have seen that the brain is most amenable to modifying a nascent script while a child is still smarting from an upsetting situation, or while painful feelings associated with a past hurtful event are present, say, for example, as a child describes a disturbing incident.

Although a knowledge of scripts can refine our natural tendencies to be better attuned and insightful with our children, it is very important for parents to remember that, as I stated in chapter 2, everyone will have scripts. They are natural to the way in which our brain is designed. As parents we cannot expect to allay all of the ways in which a child might develop a script. If we attempt to protect our children from all script-producing situations, we will be overly protective and overly anxious, and this will set the stage for the creation of amygdala scripts!

As parents we can simply do our best to aid our young children in their vulnerability, and when they are old enough we can teach them about scripts and other ways in which they can befriend their minds. But we cannot secure them from a world that is unpredictable. The most important thing that we can provide to our children is an experienced knowledge of their basic lovability or basic goodness, and this comes naturally through our love of them. Beyond that we simply do our parenting best, while practicing kindness to ourselves when we make mistakes.

SUGGESTED READINGS

Throughout the book many readings that the reader should find interesting and helpful have been cited and recommended. Here are a few that might be especially useful.

Suggestions for Therapists

Amygdala Script Theory is indebted to a long list of researchers and theoreticians that have come before. Special examples of this include Joseph LeDoux's research into the workings of the amygdala, Eye Movement Desensitization and Reprocessing (EMDR) therapy, mindfulness-based therapies, and exposure treatments. The following books offer more detailed information regarding original works that contribute to this theory along with a wealth of additional information and useful treatment suggestions.

Foa, Edna, Elizabeth Hembree, and Barbara Rothbaum. *Prolonged Exposure Therapy for PTSD: Emotional Processing of Traumatic Experiences*. New York: Oxford University Press, 2007.

Kabat-Zinn, Jon. *Full Catastrophe Living: Using the Wisdom of Your Body and Mind to Face Stress, Pain, and Illness*. New York: Delacourt, 1990.

LeDoux, Joseph E. *The Emotional Brain*. New York: Simon and Schuster, 1996.

Segal, Zindel V., J. Mark G. Williams, and John D. Teasdale. *Mindfulness-Based Cognitive Therapy for Depression: A New Approach to Preventing Relapse*. New York: Guilford Press, 2002.

Shapiro, Francine. *Eye Movement Desensitization and Reprocessing*. 2nd ed. New York: Guilford Press, 2001.

The Three-Step Practice is a tool that fits a variety of other proven approaches. Here are some of the books that I can recommend for discovery of many treatment applications that complement or are complemented by the Three-Step Practice.

Beck, A. T., A. J. Rush, and G. Emery. *Cognitive Therapy for Depression*. New York: Guilford Press, 1979. This is a rich source for the discovery of proven interventions that enhance a therapist's work with depression-related types of difficulties.

Bourne, Edmund J. *The Anxiety and Phobia Workbook*. 4th ed. Oakland, Calif.: New Harbinger Publications, 2005. This book serves similarly for the treatment of anxiety-related difficulties.

Greenberger, Dennis, and Christine Padesky. *The Clinician's Guide to Mind over Mood*. New York: Guilford Press, 1995.

Young, Jeffrey. *Cognitive Therapy for Personality Disorders: A Schema-focused Approach*. 3rd ed. Sarasota, Fla.: Professional Resource Press, 1999. This book and the previous one complement *Cognitive Therapy for Depression* in applying cognitive principles to people who suffer from Axis II types of problems—problems that I describe as mega-scripts.

Suggestions for Self-Help

Many self-help books exist, and many of them lack the merit that comes from a foundation in good research and proven techniques. Listed below are a few books that have been shown to provide benefit.

FOR ANXIETY-RELATED DIFFICULTIES

Bourne, Edmund J. *The Anxiety and Phobia Workbook*. Oakland, Calif.: New Harbinger Publications, 1990. This workbook has helped innumerable people with anxiety problems by offering a variety of proven approaches. The exercises found in this book both enhance and are enhanced by the Three-Step Practice.

Kabat-Zinn, Jon. *Full Catastrophe Living: Using the Wisdom of Your Body and Mind to Face Stress, Pain, and Illness*. New York: Delacourt, 1990. This book is supported by research and written in a very applicable style. Along with the author's tapes, these are helpful resources for anyone who is dealing with an unusual degree of stress.

FOR RECOVERY FROM TRAUMA

Rothbaum, Barbara Olasov, and Edna Foa. *Reclaiming Your Life after Rape: Cognitive-Behavioral Therapy for Post-Traumatic Stress Disorder*. New York: Oxford University Press, 2004. A very well-researched and practical guide that has helped many survivors. Although the book is directed primarily to those who are experiencing Post-Traumatic Stress Disorder, this book has suggestions that are very effective for those who have less debilitating symptoms as the result of a rape and for those who have suffered from other types of traumatic experiences.

FOR COUPLE SKILLS

Hendrix, Harville. *Getting the Love You Want*. New York: Harper Perennial, 1990. Hendrix's book offers numerous concrete suggestions that many readers have reported as helpful in their relationships.

McKay, Matthew, Patrick Fanning, and Kim Paley. *Couple Skills*. Oakland, Calif.: New Harbinger Publications, 2006. This is the kind of book that psychologists find easy to recommend. It offers many concrete exercises and strategies that have been shown to improve relationships.

FOR LOSS

Fisher, Bruce. *Rebuilding When Your Relationship Ends*. Atascadero, Calif.: Impact, 2000). This book aids not only in getting through the grief of a divorce but also in dealing with the empty aftermath of such an experience.

McWilliams, Peter, Harold H. Bloomfield, and Melba Colgrove. *How To Survive the Loss of a Love* (Santa Monica, Calif.: Prelude Press, 1973). I have recommended this book to hundreds of people, and only one of them reported that it was not helpful. An easy read and reference book that is directed to those who are looking for guidance in their grief after a breakup, but also helpful when any major loss of a love occurs.

FOR DEPRESSION

Greenberger, Dennis, and Christine Padesky, *Mind over Mood: Change How You Feel by Changing the Way You Think*. New York: Guilford Press, 1995. Written by masters in the field, *Mind over Mood* presents a cognitive and behavioral approach to treating depression.

Segal, Zindel V., J. Mark G. Williams, and John D. Teasdale. *Mindfulness-based Cognitive Therapy for Depression: A New Approach to Preventing Relapse*. New York: Guilford Press, 2001. This book describes a well-researched, groundbreaking approach that utilizes mindfulness in a manner proven to be effective for those who have recurring depression.

NOTES

Preface

1. Examples of ground-breaking literature popular among graduate students of this time include Carl Rogers, *On Becoming a Person* (Boston: Houghton Mifflin, 1961, 1995); Fritz Perls, *Ego, Hunger and Aggression* (New York: Random House, 1969); Fritz Perls, *Gestalt Therapy Verbatim* (Lafayette, Calif.: Real People's Press, 1969); Otto Kernberg, *Borderline Conditions and Pathological Narcissism* (New York: Jason Aronson, 1975); James Masterson, *Psychotherapy of the Borderline Adult* (New York: Brunner/Mazel, 1976); Elizabeth Kubler-Ross, *On Death and Dying* (New York: Simon and Schuster, 1997); and James Hillman, *The Myth of Analysis* (Evanston: Northwestern University Press, 1972).

Introduction

1. LeDoux's book *The Emotional Brain* remains a good reference for how the amygdala records painful emotion memories: Joseph LeDoux, *The Emotional Brain* (New York: Simon and Schuster, 1996). More recent reviews of relevant literature can be found in many articles, a good example being Thomas Beblo, Martin Driessen, Markus Mertens, et al., "Functional MRI Correlates of the Recall of Unresolved Life Events in Borderline Personality Disorder," *Psychological Medicine* 36 (2006): 845–856.

2. An "emotion memory" is both neurobiologically and experientially different from what people normally think of as a memory. Therefore, an "emotion memory" is distinctly different from an "emotional memory" (a historical memory that has some emotional qualities).

3. LeDoux, *Emotional Brain*, is again a good place to begin to review some of the research relevant to this. More recent examples include the article cited above, Beblo et al., "Functional MRI Correlates." Very good examples of recent research and summaries of previous studies can be found in the following articles: Katrina Carlsson, Karl Magnus Petersson, Daniel Lundqvist, Andreas Karlsson, Martin Ingvar, and Arne Öhman, "Fear and the Amygdala: Manipulation of Awareness Generates Differential Cerebral Responses to Phobic and Fear-relevant (but Nonfeared) Stimuli," *Emotion* 4 (2004): 340–353; R. J. Dolan, "Emotion, Cognition, and Behavior," *Science* 298 (2008): 1191–1194; and K. L. Phan, T. Wager, S. F. Taylor, and I. Liberzon, "Functional Neuroanatomy of Emotion: A Meta-analysis of Emotion Activation Studies in PET and fMRI," *NeuroImage* 16 (2002): 331–348.

4. A good review of research regarding the evolution of the amygdala can be found in F. Laberge, S. Muhlenbrock-Lenter, W. Grunwald, and G. Roth, "Evolution of the Amygdala: New Insights from Studies in Amphibians," *Brain Behavior Evolution* 67 (2006): 177–187.

5. As we have already noted, when an emotion memory is activated, many other regions of the brain in addition to the amygdala are activated to produce the imagery as well as the assumptions and the behaviors associated with such a memory. By focusing on the amygdala and its relationship to the neocortical regions (also a generic term that does not do justice to the complexity of functions associated with it), we are using a shorthand description. Although this is simplified, it nevertheless effectively portrays the general process that occurs in psychotherapy.

6. A partial review of the research that demonstrates the effectiveness of exposure therapy can be found in the article by Joan M. Cook, Paula P. Schnurr, and Edna B. Foa, "Bridging the Gap between Post-traumatic Stress Disorder Research and Clinical Practice: The Example of Exposure Therapy," *Psychotherapy: Theory, Research, Practice, Training* 41 (2004): 374–387. Also see Rudi De Raedt, "Does Neuroscience Hold Promise for the Further Development of Behavior Therapy? The Case of Emotional Change after Exposure in Anxiety and Depression," *Scandinavian Journal of Psychology* 47 (2006): 225–236.

7. W. Jake Jacobs and Lynn Nadel, "Neurobiology of Reconstructed Memories," *Psychology, Public Policy, and Law* 4 (1998): 1110–1134, provides a good summary of this research.

8. The techniques referred to in the book activate old painful memories in a context that promotes the conditioning of neocortical responses. These approaches have been shown to be effective and safe in numerous carefully constructed studies. There has recently been public concern that sometimes the activation of painful memories can have negative consequences. It is true that when painful memories are activated in contexts that do not include neocortical conditioning (for example, abreactive therapies) they can actually be harmful to some people. The techniques that we suggest in chapter 4 are based upon EMDR and cognitive/behavioral exposure approaches that are supported by a large body of rigorous, empirical research. Some of the research is cited above.

Chapter 1: Discovering Amygdala Scripts

1. For reasons of confidentiality, some aspects of the description of Mary have been changed.

2. See, for example, W. J. Curtis and D. Cicchetti, "Moving Research on Resilience into the 21st Century: Theoretical and Methodological Considerations in Examining the Biological Contributors to Resilience," *Development and Psychopathology* 15 (2003): 773–810.

3. For a good review of this research, see C. Holle and R. Ingram, "On the Psychological Hazards of Self-criticism," in *Self-criticism and Self-enhancement: Theory, Research, and Clinical Implications*, edited by Edward C. Chang (Washington, D.C.: American Psychological Association, 2008), 55–71. Recent original research on the relationship between habitual self-negating thoughts and depression include F. Benedetti, B. Barbini, M. C. Fulgosi, A. Pontiggia, C. Colombo, and E. Smeraldi, "Cognitive Assessment of Depression: A New Test for Mood Disorders," *Clinical*

Neuropsychiatry: Journal of Treatment Evaluation 2 (2005): 149–165; E. Calvete and J. K. Connor-Smith, "Automatic Thoughts and Psychological Symptoms: A Cross-Cultural Comparison of American and Spanish Students," *Cognitive Therapy and Research* 29 (2005): 201–217; and J. Evans, J. Heron, G. Lewis, R. Araya, and D. Wolke, "Negative Self-schemas and the Onset of Depression in Women: Longitudinal Study," *British Journal of Psychiatry* 186 (2005): 302–307. See, for example, G. Feixas, M. I. Erazo-Caicedo, S. L. Harter, and L. Bach, "Construction of Self and Others in Unipolar Depressive Disorders: A Study Using Repertory Grid Technique," *Cognitive Therapy and Research* 32 (2008): 386–400. Also see A. H. Mezulis, J. S. Hyde, and L. Y. Abramson, "The Developmental Origins of Cognitive Vulnerability to Depression: Temperament, Parenting, and Negative Life Events in Childhood as Contributors to Negative Cognitive Style," *Developmental Psychology* 42 (2006): 1012–1025.

4. The actual processes referred to in this text that are attributed directly or indirectly to the amygdala are mediated by the amygdala but also include activities of various other parts of the brain, including the anterior cingulate cortex, the orbital frontal cortex, and the insula areas of the brain. For a more complete picture of the complexities of these processes, see Carlsson et al., "Fear and the Amygdala". This greater level of complexity, however, does not change the principles that are presented in this book.

Chapter 2: Amygdala Scripts

1. Since the question is now regarded as resolved, very little research into the general efficacy of psychotherapy has occurred in the last decade or so. Seligman provides a good summary of the research into the general effectiveness of psychotherapy. See M.E.P. Seligman, "The Effectiveness of Psychotherapy: The Consumer Reports Study," *American Psychologist* 50 (1995): 965–974.

2. The issue of whether one school of psychotherapy is superior to another is complicated and remains unresolved. This is largely because the definition of what constitutes successful treatment varies, and different therapists—even those who share the same theoretical orientation—often do not employ the same tactics in their actual treatment sessions. The reader who would like to see examples of how this question has been debated might be interested in J. D. Herbert and K. T. Mueser, "The Proof Is in the Pudding: A Commentary on Persons," *American Psychologist* 46 (1991): 1347–1348; or J. S. Berman, R. C. Miller, and P. J. Massman, "Cognitive Therapy versus Systematic Desensitization: Is One Treatment Superior?" *Psychological Bulletin* 97 (1985): 451–461. More recent research has largely abandoned trying to compare various theoretical schools in favor of focusing upon the effectiveness of therapy tools that are easily adapted to a variety of therapeutic approaches. Examples of this more recent research are cited frequently throughout this book.

3. Although Kabat-Zinn began to demonstrate the salutary effects of mindfulness in the 1980s, it has only been in the last decade that the effective use of mindfulness has attracted broader scientific attention. To learn more about this, the reader is referred to Brown and Ryan's review of relevant literature: Kirk Warren Brown and Richard M. Ryan, "The Benefits of Being Present: Mindfulness and Its Role in Psychological Well-being," *Journal of Personality and Social Psychology* 84 (2003): 822–848. Another relevant article describing a study that demonstrates the

utility of mindfulness in the treatment of depression is Helen Ma and John Teas-dale, "Mindfulness-based Cognitive Therapy for Depression: Replication and Exploration of Differential Relapse Prevention Effects," *Journal of Consulting and Clinical Psychology* 72 (2004): 31–40. Some of Kabat-Zinn's work can be explored in J. Kabat-Zinn, L. Lipworth, and R. Burney, "The Clinical Use of Mindfulness Med-itation for the Self-regulation of Chronic Pain," *Journal of Behavioral Medicine* 8 (1985): 162–190; and J. Kabat-Zinn, A. O. Massion, J. Kristeller, L. G. Peterson, et al., "Effectiveness of a Meditation-based Stress Reduction Program in the Treatment of Anxiety Disorders," *American Journal of Psychiatry* 149 (1992): 936–943. For a look at a summary of his research, I recommend J. Kabat-Zinn, *Full Catastrophe Living: Using the Wisdom of Your Body and Mind to Face Stress, Pain, and Illness* (New York: Delacourt, 1990).

4. I suggest the following sampling to those readers who are interested in some of the original research showing the role that insight into the origins of psycho-logical problems plays in creating a positive therapeutic outcome: C. J. Gelso, D. M. Kivlighan Jr., B. Wine, A. Jones, and S. C. Friedman, "Transference, Insight, and the Course of Time-limited Therapy," *Journal of Counseling Psychology* 44 (1997): 209–217; P. Hoglend, V. Engelstad, O. Sorbye, et al., "The Role of Insight in Exploratory Psychodynamic Psychotherapy," *British Journal of Medical Psychology* 67 (1994): 305–317; Douglas K. Snyder, Robert M. Wills, and Arveta Grady-Fletcher, "Risks and Challenges of Long-term Psychotherapy Outcome Research: Reply to Jacobson," *Journal of Consulting and Clinical Psychology* 59 (1991): 146–149; A. Fuhri-man and G. M. Burlingame, "Consistency of Matter: A Comparative Analysis of Individual and Group Process Variables," *Counseling Psychologist* 18 (1990): 6–63; W. B. Stiles, "Measurement of the Impact of Psychotherapy Sessions," *Journal of Consulting and Clinical Psychology* 48 (1980): 176–185; Stacey E. Holmes and Dennis M. Kivlighan Jr., "Comparison of Therapeutic Factors in Group and Individual Treat-ment Processes," *Journal of Counseling Psychology* 47 (2000): 478–484; Addie Fuhri-man, Stuart Drescher, Eric Hanson, et al., "Refining the Measurement of Curativeness: An Empirical Approach," *Small Group Behavior* 17 (1986): 186–201; and Dennis M. Kivilighan Jr., Karen D. Multon, and Michael J. Patton, "Insight and Symptom Reduction in Time-limited Psychoanalytic Counseling," *Journal of Coun-seling Psychology* 47 (2000): 50–58.

5. Although a great deal of supportive research has occurred since it appeared, an early review of the effectiveness of cognitive therapy remains relevant and can be found in A. T. Beck, "Cognitive Therapy: Past, Present, and Future," *Journal of Consulting and Clinical Psychology* 61 (1993): 194–198.

6. Exposure is the behavioral tool that is used in some cases (described in chap-ter 5) in the Three-Step Method. Relevant original studies of its application are described in the following: Edna B. Foa and Michael J. Kozak, "Emotional Process-ing of Fear: Exposure to Corrective Information," *Psychological Bulletin* 99 (1986): 20–35; Peter J. Norton and Esther C. Price, "A Meta-analytic Review of Adult Cog-nitive-behavioral Treatment Outcome across the Anxiety Disorders," *Journal of Nervous and Mental Disease* 195 (2007): 521–531; and Elizabeth Brondolo, Raymond DiGiuseppe, and Raymond Chip Tafrate, "Exposure-based Treatment for Anger Problems: Focus on the Feeling," *Cognitive and Behavioral Practice* 4 (1997): 75–98.

7. This is not an exhaustive list. Other components of effective psychotherapy are noted in chapter 8.

8. Appendix B, "Neurobiology Research Notes," includes additional research that supports the basis for the following discussions.

9. See Daniel Goleman, *Emotional Intelligence* (New York: Bantam Books, 1995).

10. LeDoux was perhaps the first researcher to appreciate the significance of the robust nature of amygdala-mediated memories; J. E. LeDoux, "Indelibility of Subcortical Emotional Memories," *Journal of Cognitive Neuroscience* 1 (1989): 238–243.

11. Rejection, including repeated or especially harsh criticism and shaming, imply to individuals that their continued acceptance as members of a family, troop, or tribe is in jeopardy. A loss of group membership was potentially life-threatening in early primate history.

12. The interested reader will find a summary discussion of research relevant to the emotional nature of amygdala-mediated responses in J. E. LeDoux, "Emotion Circuits in the Brain," *Annual Review of Neuroscience* 231: 155–184.

13. LeDoux's article mentioned above also summarizes some of the early research that pointed to the unconscious nature of an activated amygdala-mediated response. Also see R. J. Dolan and Patrick Vuillemier, "Amygdala Automaticity in Emotional Processing," *Special Emphasis on the Amygdala, Annals of the New York Academy of Sciences 985* (2003): 348–355; and Phan, Wager, Taylor, and Liberzon, "Functional Neuroanatomy of Emotion," which describe more recent research regarding the unconscious nature of such responses.

14. See T. W. Buchanan, "Retrieval of Emotional Memories," *Psychological Bulletin 133* (2007): 761–779 for a recent review of relevant research.

15. A description of recent research regarding the robust nature of amygdala-mediated emotion memories can be found in Natalie L. Denburg, Tony W. Buchanan, Daniel Tranel, and Ralph Adolphs, "Evidence for Preserved Emotional Memory in Normal Older Persons," *Emotion 3* (2003): 239–253; Florian Dolcos, Kevin S. LaBar, and Roberto Cabeza, "Remembering One Year Later: Role of the Amygdala and the Medial Temporal Lobe Memory System in Retrieving Emotional Memories," *PNAS Proceedings of the National Academy of Sciences of the United States of America 102* (2005): 2626–2631.

16. Described, for example, in Denburg et al., "Preserved Emotional Memory."

17. In 1979, Aaron T. Beck's book set the standard that still applies as fundamental to cognitive-therapy treatment. See Aaron T. Beck, A. J. Rush, and G. Emery, *Cognitive Therapy of Depression* (New York: Guilford Press, 1979). More recent examples of mainstream treatment approaches in cognitive therapy can be found in Jeffrey E. Young, *Cognitive Therapy for Personality Disorders: A Schema-focused Approach*, 3rd ed. (Sarasota, Fla.: Professional Resource Press, 1999); and Christine Padesky, "Schema Change Processes in Cognitive Therapy," *Clinical Psychology and Psychotherapy 1* (1994): 267–278. I recommend all of these publications highly to all who are interested in cognitive therapy, even though their focus is upon belief change outside the context of a person's internal narrative stream.

18. There is a good summary of these different views in M. J. Power, "Cognitive Therapy: An Outline of Theory, Practice and Problems," *British Journal of Psychotherapy 5* (1989): 544–556.

19. There is a great deal of research on the relationship between painful events that occurred in childhood and later psychological problems. Typical of this type of

research is the work of two Canadian researchers, Kathryn Hildyard and David Wolfe, who were able to show the relationship between childhood neglect and a variety of later psychological problems. See Kathryn L. Hildyard and David A. Wolfe, "Child Neglect: Developmental Issues and Outcomes," *Child Abuse and Neglect* 26 (2002): 679–695. Brandon Gibb and fellow researchers were able to distinguish certain types of early childhood trauma associated with depression from those that seemed associated with anxiety. See B. E. Gibb, A. C. Butler, and J. S. Beck, "Childhood Abuse, Depression, and Anxiety in Adult Psychiatric Outpatients," *Depression and Anxiety* 17 (2003): 226–228.

Painful events that create scripts are not always the result of bad parenting, but in many cases parental mistakes produce later psychological difficulties. Dante Cicchetti has spent his life researching how parental maltreatment creates the basis for many psychological problems, and his 2004 article provides a good summary of his work and that of others: Dante Cicchetti, "An Odyssey of Discovery: Lessons Learned through Three Decades of Research on Child Maltreatment," *American Psychologist* 59 (2004): 731–741.

Some researchers have been able to identify the neurobiological substrates that underlie painful childhood experiences. For example, a couple of researchers at Emory University, Christine Heim and Charles Nemeroff, did an interesting study and were able to tie painful childhood experiences to neurobiological correlates of anxiety and depression. See C. Heim and C. B. Nemeroff, "The Role of Childhood Trauma in the Neurobiology of Mood and Anxiety Disorders: Preclinical and Clinical Studies," *Biological Psychiatry* 49 (2001): 1023–1039. Another example of this type of research can be found in a study by Linda Luecken and Kathy Lemery, who have been studying the biological pathways through which early stressful events in childhood can result in adulthood predispositions for vulnerability to stress disorders. See L. J. Luecken and K. S. Lemery, "Early Caregiving and Physiological Stress Responses," *Clinical Psychology Review* 24 (2004): 171–191.

Chapter 3: Overview of the Three-Step Practice for Mastering Amygdala Scripts

1. Although awareness of emotions has long been valued by psychotherapists, studies such as those done by researchers Kivilighan Jr., Multon, and Brossart served to empirically validate and refine emotional-awareness tools. See Dennis M. Kivilighan Jr., Karen D. Multon, and D. Brossart, "Helpful Impacts in Group Counseling: Development of a Multidimensional Rating System," *Journal of Counseling Psychology* 43 (1996): 347–355. A more recent example of this type of research is found in a comprehensive review of empirical research regarding mindfulness in Brown and Ryan, "Benefits of Being Present"; and another example of more recent, relevant research can be found in Kivilighan Jr., Multon, and Patton, "Insight and Symptom Reduction."

2. Beginning with Freud, psychotherapists have utilized insight into the personal, historical origins of a psychological problem as an important tool in the treatment of psychological problems. Neurobiological research has recently been able to point to some of the neurological mechanisms that underlie this particular use of insight and provide likely explanations, in terms of brain science, as to why such insight is effective. Excellent examples of this can be found in E. A. Phelps, "Human Emotion and Memory: Interactions of the Amygdala and Hippocampal Complex," *Current Opinion in Neurobiology* 14 (2004): 198–202; De Raedt, "Does

Neuroscience Hold Promise?"; and R. J. Davidson, D. C. Jackson, and N. H. Kalin, "Emotion, Plasticity, Context, and Regulation: Perspectives from Affective Neuroscience," *Psychological Bulletin* 126 (2000): 890–909.

3. Quotation from Gertrude Sly, who raised eight children in a two-bedroom house.

Chapter 4: Step 1

1. Research by Critchley and colleagues provides a good summary of the relationship between the brain and those subtle physical sensations that we recognize as emotions. See H. D. Critchley, S. Wiens, P. Rotshtein, A. Öhman, and R. J. Dolan, "Neural Systems Supporting Interoceptive Awareness," *Nature Neuroscience* 7 (2004): 189–195. Pollatos and colleagues also provide a very good summary of these mechanisms. See O. Pollatos, K. Gramann, and R. Schandry, "Neural Systems Connecting Interoceptive Awareness and Feelings," *Human Brain Mapping* 28 (2007): 9–18.

2. By consciousness, we mean human consciousness, whereby humans experience themselves as semi-independent actors in the world, able to observe and know that they are observing a variety of phenomena, including their own internal states.

3. The anterior insula plays a central role in this process but acts in concert with many other areas of the brain, notably the anterior cingulate cortex, the orbito frontal cortex, and the periaqueductal gray.

4. In the appendices there is a more detailed review of neuroscience relevant to this, and the reader who would like an even more in-depth discussion of these mechanisms will probably be interested in the studies and summaries of research by Carlsson, Lane, and colleagues. See Carlsson et al., "Fear and the Amygdala"; and Richard D. Lane, Gereon R. Fink, Phyllis M-L. Chau, and Raymond J. Dolan, "Neural Activation during Selective Attention to Subjective Emotional Responses," *Neuroreport: An International Journal for the Rapid Communication of Research in Neuroscience* 8 (1997): 3969–3972.

5. For a good example of how emotional insight can be used in therapy, see L. S. Greenberg and S. H. Warwar, "Homework in an Emotion-focused Approach to Experiential Therapy," *Journal of Psychotherapy Integration* 16 (2006): 178–200.

6. Psychologists have studied how the development of a sense of self occurs. See, for example, "The Meaning and Measurement of Ego Development" by J. Loevinger in *American Psychologist* 21 (1966): 195–206.

7. The Asian world does not make such a strong distinction. In India and Tibet, for example, the mind is considered to be in the heart, experienced in part through a variety of physical sensations that are understood to be a manifestation of the mind.

8. See, for example, Gordon Baker and Katherine J. Morris, *Descartes' Dualism* (London: Routledge, 1996).

9. Although body-oriented psychotherapies have existed for many years, they have been consistently marginalized by mainstream psychotherapy, largely because the techniques that were previously promoted have not proven themselves in the arena of empirical research.

10. See, for example, Kabat-Zinn, *Full Catastrophe Living*.

11. The other two insights are that a great deal of what determines how we feel and behave operates with little or no conscious awareness, and that we are much more profoundly affected by our histories than was ever imagined before Freud.

12. The brief discussion offered here regarding mindfulness does not encompass the subtler practices and understanding of mindfulness that have been developed in Buddhism.

13. Kabat-Zinn, *Full Catastrophe Living*, provides an excellent summary of some of the early research into the effectiveness of mindfulness as a tool for Western psychology.

14. Kabat-Zinn's 1985 study is a classic example of this early research. See Kabat-Zinn, Lipworth, and Burney, "Clinical Use of Mindfulness Meditation." It was Linehan's book, *Cognitive-behavioral Treatment of Borderline Personality Disorder* (New York: Guilford Press, 1993), that brought her to the attention of the psychological world. The ensuing research that replicated her results cemented her reputation and the therapeutic tools she had developed. More recently, Segal, Williams, and Teasdale published a very influential book demonstrating how mindfulness practices can play a central in the treatment of chronic depression: Z. V. Segal, J.M.G. Williams, and J. D. Teasdale, *Mindfulness-based Cognitive Terapy for Depression—A New Approach to Preventing Relapse* (New York: Guilford Press, 2002). Their approach, like that of Linehan and Kabat-Zinn, is supported by strong and well-designed research.

15. Examples of this early research can be found in: J. Kabat-Zinn, A. O. Massion, J. Kristeller, et al., "Effectiveness of a Meditation-based Stress Reduction Program in the Treatment of Anxiety Disorders," *American Journal of Psychiatry* 149 (1992): 936–943; Kabat-Zinn, Lipworth, and Burney, "Clinical Use of Mindfulness Meditation"; and John J. Miller, Ken Fletcher, and Jon Kabat-Zinn, "Three-year Follow-up and Clinical Implications of a Mindfulness Meditation-based Stress Reduction Intervention in the Treatment of Anxiety Disorders," *General Hospital Psychiatry* 17 (1995): 192–200.

16. Robins describes some of the difficulties in treating this population—usually diagnosed as having "borderline personality disorders"—and how they might be addressed in his article: Clive J. Robins and Alex L. Chapman, "Dialectical Behavior Therapy: Current Status, Recent Developments, and Future Directions," *Journal of Personality Disorders* 18 (2004): 73–89.

17. Lynch et al. note the role that this aspect of Linehan's treatment plays and its effectiveness in their 2006 study: Thomas R. Lynch, Alexander L. Chapman, M. Zachary Rosenthal, Janice R. Kuo, and Marsha M. Linehan, "Mechanisms of Change in Dialectical Behavior Therapy: 2004 Theoretical and Empirical Observations," *Journal of Clinical Psychology* 62 (2006): 459–480. See Robins and Chapman, "Dialectical Behavior Therapy."

18. See an excellent example of this research in Shawn P. Cahill, Maureen H. Carrigan, and Christopher B. Frueh, "Does EMDR Work? And If So, Why? A Critical Review of Controlled Outcome and Dismantling Research," *Journal of Anxiety Disorders* 13 (1999): 5–33.

19. An example of this can be found in Francine Shapiro, *Eye Movement Desensitization and Reprocessing,* 2nd ed. (New York: Guilford Press, 2001), 57–64.

20. Francine Shapiro invites well-known researchers and therapists to comment on the role that her EMDR therapy has played in their work. See Francine Shapiro, ed., *EMDR as an Integrative Psychotherapy Approach* (Washington D.C.: American Psychological Association, 2002).

21. It was in 1998 that Edna Foa and her colleagues published a research paper highlighting this aspect of exposure therapy. See Lisa H. Jaycox, Edna B. Foa, and Andrew R. Morral, "Influence of Emotional Engagement and Habituation on Exposure Therapy for PTSD," *Journal of Consulting and Clinical Psychology* 66 (1998): 185–192. Exposure therapies are now commonly used to treat a wide range of problems. Our use of this treatment tool will be described in more detail in chapter 5.

22. Jasper Smits and colleagues have contributed to this research. See Jasper A. J. Smits, Mark B. Powers, Yongrae Cho, and Michael J. Telch, "Mechanism of Change in Cognitive-behavioral Treatment of Panic Disorder: Evidence for the Fear of Fear Mediational Hypothesis," *Journal of Consulting and Clinical Psychology* 72 (2004): 646–652, which provides an interesting summary of one line of thinking regarding how attunement to the physical sensations aids in the treatment of anxiety disorders. Another approach suggests that the identification and naming of negative emotional states seems to have an inherently salutary effect by activating a moderating neocortical response. For an example of this research and reasoning, see Matthew D. Lieberman, Naomi I. Eisenberg, Molly J. Crockett, Sabrina M. Tom, Jennifer H. Pfeifer, and Baldwin M. Way, "Putting Feelings into Words: Affect Labeling Disrupts Amygdala Activity in Response to Affective Stimuli," *Psychological Science* 18 (2007): 421–428.

23. Brown and Ryan offer a comprehensive summary of the pre-2003 research regarding the value and mechanisms of mindfulness practices demonstrated in Western empirical research. See Brown and Ryan, "Benefits of Being Present."

24. After years of research demonstrating the effectiveness of exposure and cognitive therapy, Foa and her colleagues have begun to dismantle these tools, separating out the cognitive aspects from other aspects to identify the major contributors to its effectiveness. To their surprise they have found that exposure therapy, without an explicit focus on changing cognitive beliefs, is just as effective as exposure therapy with the addition of cognitive therapy. See Edna B. Foa, E. A. Hembree, S. P. Cahill, et al., "Randomized Trial of Prolonged Exposure for Posttraumatic Stress Disorder with and without Cognitive Restructuring: Outcome at Academic and Community Clinics," *Journal of Consulting and Clinical Psychology* 73 (2005): 953–964. Teasdale and colleagues discovered similar results and offered a comprehensive explanation that is found in the present volume: J. D. Teasdale, R. G. Moore, H. Hayhurst, M. Pope, S. Williams, and Z. V. Segal, "Metacognitive Awareness and Prevention of Relapse in Depression: Empirical Evidence," *Journal of Consulting and Clinical Psychology* 70 (2002): 275–287.

25. A more comprehensive discussion of this theory and a summary of the research that supports it can be found in Teasdale et al., "Metacognitive Awareness."

26. Linehan, *Cognitive-behavioral Treatment of Borderline Personality Disorder* offers a detailed account of how therapists and clients can apply mindful awareness.

27. Kennedy-Moore reviews some of this research and provides guidelines for the ways in which identifying emotions can be most useful. See Eileen Kennedy-Moore and Jeanne C. Watson, "How and When Does Emotional Expression Help?" *Review of General Psychology* 5 (2001): 187–212.

Chapter 5: Step 2

1. See Carlsson et al., "Fear and the Amygdala."

2. LeDoux, *Emotional Brain*.

3. See Jeffrey B. Rosen and Melanie P. Donley, "Animal Studies of Amygdala Function in Fear and Uncertainty: Relevance to Human Research," *Biological Psychology* 73 (2006): 49–60; Michael B. Keeley, Marcelo A. Wood, Carolina Isiegas, Joel Stein, Kevin Hellman, Hannenhalli Sridhar, et al., "Differential Transcriptional Response to Nonassociative and Associative Components of Classical Fear Conditioning in the Amygdala and Hippocampus," *Learning and Memory* 13 (2006): 135–142; Linda M. Rorick-Kehn and Joseph E. Steinmetz, "Amygdala Unit Activity during Three Learning Tasks: Eyeblink Classical Conditioning, Pavlovian Fear Conditioning, and Signaled Avoidance Conditioning," *Behavioral Neuroscience* 119 (2005): 1254–1276; and A. Bechara, D. Tranel, H. Damasio, R. Adolphs, C. Rockland, and A. R. Damasio, "Double Dissociation of Conditioning and Declarative Knowledge Relative to the Amygdala and Hippocampus in Humans," *Science* 269 (August 25, 1995): 1115–1118.

4. The ethics of hurting animals (even without injury) is questionable and widely debated.

5. As mentioned in chapter 1, the term "amygdala script" seems especially descriptive from the standpoint of problematic human behavior. The terms "emotion memory" and "gist" memories (as distinct from declarative memories) and certain subsets of "conditioned responses" are all terms that have been used to describe phenomena that we refer to as "amygdala scripts."

6. See, for example, R. J. Dolan, "Emotion, Cognition, and Behavior," *Science* 298 (November 8, 2002): 1191–1194.

7. See, for example, Ralph Adolphs, Natalie L. Denburg, and Daniel Tranel, "The Amygdala's Role in Long-term Declarative Memory for Gist and Detail," *Behavioral Neuroscience* 115 (2001): 983–992; and Carlsson et al., "Fear and the Amygdala."

8. The recording of declarative memories in the brain is a process mediated by the hippocampal regions of the brain, as compared to the emotion memories (scripts) mediated by the amygdala. The hippocampal regions may, however, also play a subsidiary role in the automatic recall of the context associated with an emotion memory.

9. Denburg et al., "Preserved Emotional Memory."

10. Ibid.

11. Research suggests that shyness is an example of scripts that combine a genetic predisposition, perhaps toward tending to be introverted, with painful experiences that lead to an amygdala script. Children who get a clear and repeated message that their introverted tendencies are normal and acceptable—and that such tendencies even have potential benefits—are less likely to develop an amygdala script associated with shyness, since scripts are dependent on painful experiences.

12. A growing body of research suggests that diagnostic categories may some day be set aside in favor of more useful and accurate means of categorizing psychological difficulties. See, for example, D. Watson, "Rethinking the Mood and

Anxiety Disorders: A Quantitative Hierarchical Model for DSM-V," *Journal of Abnormal Psychology* 114 (2005): 522–536; and Lee Anna Clark, "Temperament as a Unifying Basis for Personality and Psychopathology," *Journal of Abnormal Psychology* 114 (2005): 505–521.

13. See, for example, C. M. Warner, C. H. Warner, J. Breitbach, J. Rachal, T. Matuszak, and T. A. Grieger, "Depression in Entry-level Military Personnel," *Military Medicine* 172 (2007): 795–799; K. S. Kendler, C. O. Gardner, and C. A. Prescott, "Toward a Comprehensive Developmental Model for Major Depression in Women," *American Journal of Psychiatry* 159 (2002): 1133–1145; and K. S. Kendler and C. O. Gardner, "Monozygotic Twins Discordant for Major Depression: A Preliminary Exploration of the Role of Environmental Experiences in the Aetiology and Course of Illness," *Psychological Medicine* 31 (2001): 411–423.

14. Note that I am not suggesting that because scripts are learned they do not have biological correlates in the brain. Clearly they do. Nor am I suggesting that the amygdala, when activating a script that manifests as anxiety, will activate exactly the same biological response as in the case of depression. I am suggesting that most of what we experience as depression or anxiety (for example) is learned, and that what is learned can be unlearned. This does not preclude the possibility that significant actual structural changes in the brain can be the result of intense, long-established, frequently activated scripts. I do suggest that the neocortex is functional even in those more extreme cases, and that new and different learning can occur. This is borne out by a large body of clinical research.

15. See Shapiro, *Eye Movement Desensitization and Reprocessing* (Shapiro does not use the term "scripts").

16. Shapiro's original formulation of this process included ongoing "bilateral stimulation," but subsequent research indicated that little benefit accrued from this part of the procedure, and so a number of years ago I jettisoned the bilateral stimulation part and found no reduction in overall effectiveness. See Ulrike Feske and Alan J. Goldstein, "Eye Movement Desensitization and Reprocessing Treatment for Panic Disorder: A Controlled Outcome and Partial Dismantling Study," *Journal of Consulting and Clinical Psychology* 65 (1997): 1026–1035; G. Renfrey and C. R. Spates, "Eye Movement Desensitization: A Partial Dismantling Study," *Journal of Behavior Therapy and Experimental Psychiatry* 25 (1994): 231–239; and Cahill, Carrigan, and Frueh, "Does EMDR Work?"

17. Shapiro, *Eye Movement Desensitization and Reprocessing.*

18. See Denburg et al., "Preserved Emotional Memory." Although Denburg demonstrated this in a laboratory setting, there are many common examples of this: most people who witnessed the destruction of the World Trade Center on television are left with a vivid image of that event and can also tell you where they were when they first heard about it; similarly most people who were alive when John F. Kennedy was assassinated can tell you exactly where they were and what they were doing when they first heard about it.

19. Adolphs, Denburg, and Tranel, "Amygdala's Role in Long-term Declarative Memory."

20. For a review of this research, see E. F. Loftus and D. Davis, "Recovered Memories," *Annual Review of Clinical Psychology* 2 (2006): 469–498.

21. As is the case with all seed images, we are not assuming that this happened in exactly the way he "remembered" it. Although it can be worked with as if it were a true event, its most important significance lies in that it epitomizes how Jed experienced his parents' reaction to his fear.

22. We could not do this in chapter 4 because the exposure process for reducing emotional intensity begins with accessing a seed image.

23. We might wonder why scripts are so stubbornly persistent even though they are repeatedly activated in environments where there are no painful consequences. This is because humans are very adept at suppressing and otherwise avoiding the painful feelings associated with a script, and extinction can only occur when we repeatedly experience what once seemed like an unbearable emotion. See, for example, S. T. Higgins and E. K. Morris, "Generality of Free-operant Avoidance Conditioning to Human Behavior," *Psychological Bulletin* 96 (1984): 247–272.

24. See, for example, Joseph Wolpe, *Psychotherapy by Reciprocal Inhibition* (Stanford: Stanford University Press, 1958).

25. See, for example, Joseph Wolpe, *Life without Fear: Anxiety and Its Cure* (Oakland, Calif.: Harbingers, 1988).

26. See Foa and Kozak, "Emotional Processing of Fear."

27. Cook, Schnurr, and Foa, "Bridging the Gap."

28. See Edna B. Foa, Barbara Olasov Rothbaum, David S. Riggs, and Tamera B. Murdock, "Treatment of Post-traumatic Stress Disorder in Rape Victims: A Comparison between Cognitive-behavioral Procedures and Counseling," *Journal of Consulting and Clinical Psychology* 59 (1991): 715-723; Christopher B. Frueh, Samuel M. Turner, and Deborah C. Beidel, "Exposure Therapy for Combat-related PTSD: A Critical Review," *Clinical Psychology Review* 15 (1995): 799–817; and Jeffrey E. Hecker, "Emotional Processing in the Treatment of Simple Phobia: A Comparison of Imaginal and in Vivo Exposure," *Behavioural Psychotherapy* 18 (1990): 21–34.

29. Shapiro, *Eye Movement Desensitization and Reprocessing*. Shapiro does not often use the term exposure therapy when she is describing her procedure, although it follows a protocol easily recognized as an exposure protocol.

30. Shapiro, *EMDR as an Integrative Psychotherapy Approach*.

Chapter 6: Step 3

1. Examples of endeavors to integrate cognitive therapy and other therapeutic orientations can be found in the following articles: Louis G. Castonguay, Alexander J. Schut, Deane E. Aikens, et al., "Integrative Cognitive Therapy for Depression: A Preliminary Investigation," *Journal of Psychotherapy Integration* 14 (2004): 4–20; Gillian E. Hardy, Jane Cahill, David A. Shapiro, Michael Barkham, Anne Rees, and Norman Macaskill, "Client Interpersonal and Cognitive Styles as Predictors of Response to Time-limited Cognitive Therapy for Depression," *Journal of Consulting and Clinical Psychology* 69 (2001): 841–845; Adele M. Hayes and Jennifer L. Strauss, "Dynamic Systems Theory as a Paradigm for the Study of Change in Psychotherapy: An Application to Cognitive Therapy for Depression," *Journal of Consulting and Clinical Psychology* 66 (1998): 939–947; Robert Elliott, David A. Shapiro, Jenny Firth-Cozens, et al., "Comprehensive Process Analysis of Insight Events in Cognitive-

Behavioral and Psychodynamic-Interpersonal Psychotherapies," *Journal of Counseling Psychology* 41 (1994): 449–463.

2. Beck provides a good early summary of this research: Beck, Rush, and Emery, *Cognitive Therapy of Depression.*

3. Teasdale et al., "Metacognitive Awareness," 275

4. Ibid., 284. In research parlance, watching one's thoughts without being affected by them is referred to as "metacognition." A summary of additional research that supports metacognition as a central mechanism for positive psychological change can be found in Rael Cahn and John Polich, "Meditation States and Traits: EEG, ERP and Neuroimaging Studies," *Psychological Bulletin* 132 (2006): 180–211. Other summaries of relevant research can be found in Beevers et al., who describe their research, which suggests that one's relationship with negative thoughts rather than the modification of negative thoughts results in reduced psychological symptoms. See Christopher G. Beevers and Ivan W. Miller, "Unlinking Negative Cognition and Symptoms of Depression: Evidence of a Specific Treatment Effect for Cognitive Therapy," *Journal of Consulting and Clinical Psychology* 73 (2005): 68–77.

5. Research has borne out the importance that these kinds of interactions play in the healthy development of children. See, for example, Ronald P. Rohner and Robert A. Veneziano, "The Importance of Father Love: History and Contemporary Evidence," *Review of General Psychology* 5 (2001): 382–405. Also see Theodore Dix, Elizabeth T. Gershoff, Leah N. Meunier, and Pamela C. Miller, "The Affective Structure of Supportive Parenting: Depressive Symptoms, Immediate Emotions, and Child-oriented Motivation," *Developmental Psychology* 40 (2004): 1212–1227.

6. Examples of research on the role that empathy and warmth play can be found in: Barry A. Farber and Jodie S. Lane, "Positive Regard," *Psychotherapy: Theory, Research, Practice, Training* 38 (2001): 390–395; Howard Kirschenbaum and April Jourdan, "The Current Status of Carl Rogers and the Person-centered Approach," *Psychotherapy: Theory, Research, Practice, Training* 42 (2005): 37–51; Leslie S. Greenberg, Jeanne C. Watson, Robert Elliot, and Arthur C. Bohart, "Empathy," *Psychotherapy: Theory, Research, Practice, Training* 38 (2001): 380–384; and Michael J. Lambert and Dean E. Barley, "Research Summary on the Therapeutic Relationship and Psychotherapy Outcome," *Psychotherapy: Theory, Research, Practice, Training* 38 (2001): 357–361.

7. The following are two examples of research that underline the importance of warmth and empathy in cognitive therapy: L. G. Castonguay, M. R. Goldfried, S. Wiser, P. J. Raue, and A. M. Hayes, "Predicting the Effect of Cognitive Therapy for Depression: A Study of Unique and Common Factors," *Journal of Consulting and Clinical Psychology* 65 (1996): 497–504; and J. D. Safran and L. K. Wallner, "The Relative Predictive Validity of Two Therapeutic Alliance Measures in Cognitive Therapy," *Psychological Assessment* 31 (1991): 188–195.

8. M. J. Lambert, "Psychotherapy Outcome Research: Implications for Integrative and Eclectic Therapies," in *Handbook of Psychotherapy Integration*, edited by J. C. Norcross and M. R. Goldfried (New York: Basic Books, 1992), 94–129.

9. Examples of original research that demonstrate this can be found in: B. Singer, E. Friedman, T. Seeman, G. A. Fava, and C. D. Ryff, "Protective Environments and

Health Status: Cross-talk between Human and Animal Studies," *Neurobiology of Aging* 26 (2005): S113–S118; S. E. Taylor, "Tend and Befriend: Biobehavioral Bases of Affiliation under Stress," *Current Directions in Psychological Science* 15 (2006): 273–277; K. M. Grewen, S. S. Girdler, J. Amico, and K. C. Light, "Effects of Partner Support on Resting Oxytocin, Cortisol, Norepinephrine, and Blood Pressure before and after Warm Partner Contact," *Psychosomatic Medicine* 67 (2005): 531–538; Peter Kirsch, Christine Esslinger, Qiang Chen, Daniela Mier, Stefanie Lis, Sarina Siddhanti, et al., "Oxytocin Modulates Neural Circuitry for Social Cognition and Fear in Humans," *Journal of Neuroscience* 25 (2005): 11489–11493; Paul J. Zak, Robert Kurzban, and William T. Matzner, "Oxytocin Is Associated with Human Trustworthiness," *Hormones and Behavior* 48 (2005): 522–527.

10. See, for example, Taylor, "Tend and Befriend"; Grewen et al., "Partner Support"; Singer et al., "Protective Environments"; K. C. Light, K. M. Grewen, and J. A. Amico, "More Frequent Partner Hugs and Higher Oxytocin Levels Are Linked to Lower Blood Pressure and Heart Rate in Premenopausal Women," *Biological Psychology* 69 (2005): 5–21; and Tobias Esch and George B. Stefano, "The Neurobiology of Love," *Neuroendocrinology Letters* 26 (2007): 1–18.

11. For studies and research summaries that demonstrate the role that these hormones play in addiction, see, for example, M. J. Kreek, S. D. Schlussman, G. Bart, K. S. La Forge, and E. R. Butelman, "Evolving Perspectives on Neurobiological Research on the Addictions: Celebration of the 30th Anniversary of NIDA," *Neuropharmacology* 47 (2004): 324–344; Robert L. DuPont, "Addiction: A New Paradigm," *Bulletin of the Menninger Clinic* 62 (1998): 231–242; and Richard A. Depue and Jeannine V. Morrone-Strupinsky, "A Neurobehavioral Model of Affiliative Bonding: Implications for Conceptualizing a Human Trait of Affiliation," *Behavioral and Brain Sciences* 28 (2005): 313–395.

12. See, for example, Kirsch et al., "Oxytocin Modulates Neural Circuitry."

13. Farber and Lane, "Positive Regard"; Kirschenbaum and Jourdan, "Carl Rogers"; and Greenberg et al., "Empathy."

14. Farber and Lane, "Positive Regard"; Kirschenbaum and Jourdan, "Carl Rogers"; and Greenberg et al., "Empathy."

15. Personification is not the only means that psychotherapy has found for doing this. In Freudian-based psychotherapy, this occurs when the patient "introjects" the caring concern of their therapist into his or her own psyche.

16. Perls, *Gestalt Therapy Verbatim* remains a classic in describing ways in which personification can contribute to therapy.

17. James Hillman, *Revisioning Psychology* (New York: Harper and Row, 1975).

18. Esch and Stefano, "Neurobiology of Love."

19. Singer et al., "Protective Environments"; Taylor, "Tend and Befriend"; Kirsch et al., "Oxytocin Modulates Neural Circuitry"; and Zak, Kurzban, and Matzner, "Oxytocin Is Associated with Human Trustworthiness."

20. J. Burgsdorf and J. Panksepp, "The Neurobiology of Positive Emotions," *Neuroscience and Biobehavioral Reviews* 30 (2006): 173–187. Also see G. T. Panagis, "Biopsychology of Reinforcement: Intracranial Self-stimulation Studies and the Role of Dopamine," *Psychology: The Journal of the Hellenic Psychological Society* 9 (2002): 92–115.

Chapter 7: Putting It All Together

1. V. E. Wilson and E. Peper, "The Effects of Upright and Slumped Postures on the Recall of Positive and Negative Thoughts," *Applied Psychophysiology and Biofeedback* 29 (2004): 189–195.

2. See, for example, Brown and Ryan, "Benefits of Being Present."

3. For a summary of research that portrays the benefits of mindfulness, see ibid.

Chapter 8: Therapist to Therapist

1. See D. J. Martin, J. P. Garske, and M. K. Davis, "Relation of the Therapeutic Alliance with Outcome and Other Variables: A Meta-analytical Review," *Journal of Consulting and Clinical Psychology* 68 (2000): 438–450; and S. A. Baldwin, B. E. Wampold, and Z. E. Imel, "Untangling the Alliance-outcome Correlation: Exploring the Relative Importance of Therapist and Patient Variability in the Alliance," *Journal of Consulting and Clinical Psychology* 75 (2007): 842–852.

2. E. Bordin, "The Generalizability of the Psychoanalytic Concept of the Working Alliance," *Psychotherapy: Theory, Research, and Practice* 16 (1979): 252–260.

3. See, for example, M. A. Blais, M. J. Hilsenroth, and F. D. Castlebury, "Psychometric Characteristics of the Cluster B Personality Disorders under DSM-III-R and DSM-IV," *Journal of Personality Disorders* 11 (1997): 270–278; and C. A. Sanislow, L. C. Morey, C. M. Grilo, et al., "Confirmatory Factor Analysis of DSM-IV Borderline, Schizotypal, Avoidant, and Obsessive-compulsive Personality Disorders: Findings from the Collaborative Longitudinal Personality Disorders Study," *Acta Psychiatrica Scandinavica* 105 (2002): 28–36.

4. See, for example, D. L. Herbert, R. O. Nelson, and J. D. Herbert, "Effects of Psychodiagnostic Labels, Depression Severity, and Instructions on Assessment," *Professional Psychology: Research and Practice* 19 (1988): 496–502. Also see Kristie Madsen and Peter Leech, *The Ethics of Labeling in Mental Health* (Jefferson, N.C.: MacFarland, 2007).

5. For research overview on depression, see Tori DeAngelis, "When Do Meds Make the Difference?" *Monitor on Psychology American Psychological Association* 39 (2008): 48. For some of the original research, see L. L. Hawley, M. R. Ho, D. C. Zuroff, and S. J. Blatt, "Stress Reactivity Following Brief Treatment for Depression: Differential Effects of Psychotherapy and Medication," *Journal of Consulting and Clinical Psychology* 75 (2007): 244–256; E. J. Ludman, G. E. Simon, S. Tutty, and M. Von Korff, "A Randomized Trial of Telephone Psychotherapy and Pharmacotherapy for Depression: Continuation and Durability of Effects," *Journal of Consulting and Clinical Psychology* 75 (2007): 257–266; I. M. Blackburn, S. Bishop, A. I. M. Glen, L. J. Whalley, and J. E. Christie, "The Efficacy of Cognitive Therapy in Depression: A Treatment Trial Using Cognitive Therapy and Pharmacotherapy, Each Alone and in Combination," *British Journal of Psychiatry* 139 (1981): 181–189. Also see I. Elkin, M. T. Shea, J. T. Watkins, et al., "National Institute of Mental Health Treatment of Depression Collaborative Research Program: General Effectiveness of Treatments," *Archives of General Psychiatry* 46 (1989): 971–982; and Y. Leykin, J. D. Amsterdam, R. J. DeRubeis, R. Gallop, R. C. Shelton, and S. D. Hollon, "Progressive Resistance to a Selective Serotonin Reuptake Inhibitor but Not to Cognitive Therapy in the Treatment of Major Depression," *Journal of Consulting and Clinical Psychology* 75 (2007): 267–276.

For research overview on anxiety, see DeAngelis, "When Do Meds Make the Difference?"; and for more original research, see J. Siev and D. L. Chambless, "Specificity of Treatment Effects: Cognitive Therapy and Relaxation for Generalized Anxiety and Panic Disorders," *Journal of Consulting and Clinical Psychology* 75 (2007): 513–522; L. B. Feldman and R. A. Rivas-Vazquez, "Assessment and Treatment of Social Anxiety Disorder," *Professional Psychology: Research and Practice* 34 (2003): 396–405; Michel J. Dugas, Robert Ladouceur, Eliane Léger, et al., "Group Cognitive-Behavioral Therapy for Generalized Anxiety Disorder: Treatment Outcome and Long-Term Follow-up," *Journal of Consulting and Clinical Psychology* 71 (2003): 821–825; R. Ladouceur, M. J. Dugas, M. H. Freeston, E. Léger, F. Gagnon, and N. Thibodeau, "Efficacy of a Cognitive-Behavioral Treatment for Generalized Anxiety Disorder: Evaluation in a Controlled Clinical Trial," *Journal of Consulting and Clinical Psychology* 68 (2000): 957–964.

For research overview on obsessive-compulsive disorders, see DeAngelis, "When Do Meds Make the Difference?" For some of the original research, see H. Hiss, E. B. Foa, and M. J. Kozak, "Relapse Prevention Program for Treatment of Obsessive-compulsive Disorder," *Journal of Consulting and Clinical Psychology* 62 (1994): 801–808; and Martin E. Franklin, Jonathan S. Abramowitz, Michael J. Kozak, Jill T. Levitt, and Edna B. Foa, "Effectiveness of Exposure and Ritual Prevention for Obsessive-Compulsive Disorder: Randomized Compared with Nonrandomized Samples," *Journal of Consulting and Clinical Psychology* 68 (2000): 594–602.

6. See L. A. Clark, W. J. Livesley, and L. Morey, "Personality Disorder Assessment: The Challenge of Construct Validity," *Journal of Personality Disorders* 11 (1997): 205–231; E. Sheets and W. E. Craighead, "Toward an Empirically-based Classification of Personality Pathology," *Clinical Psychology: Science and Practice* 14 (2007): 77–93. Also see T. A. Widiger and L. A. Clark, "Toward DSM-V and the Classification of Psychopathology," *Psychological Bulletin* 126 (2000): 946–963; G. Dimaggio and J. C. Norcross, "Treating Patients with Two or More Personality Disorders: An Introduction," *Journal of Clinical Psychology* 64 (2008): 127–138; and T. W. Miller, J. T. Nigg, and S. V. Faraone, "Axis I and II Comorbidity in Adults with ADHD," *Journal of Abnormal Psychology* 116 (2007): 519–528.

7. Clark, "Temperament as Unifying Basis for Personality and Psychopathology."

8. For research on personality traits remaining stable after successful treatment, see Robert R. McCrae, Paul T. Costa Jr., Fritz Ostendorf, et al., "Nature over Nurture: Temperament, Personality, and Life Span Development," *Journal of Personality and Social Psychology* 78 (2000): 173–186.

For research suggesting that these factors measure predispositions to particular types of problems, but not causes of psychological problems, see Filip De Fruyt, Karla Van Leeuwen, R. Michael Bagby, Jean-Pierre Rolland, and Frédéric Rouillon, "Assessing and Interpreting Personality Change and Continuity in Patients Treated for Major Depression," *Psychological Assessment* 18 (2006): 71–80; and McCrae et al., "Nature over Nurture."

9. For example, certain types of schizophrenia, the more extreme cases of bipolar disorder, and atypical reaction to alcohol are a few examples in which genetic or other biological loading seem to create psychological dysfunction with minimal or no environmental input. The result is that psychotherapeutic treatment aids a client in learning to manage symptoms rather than completely master them.

See G. Groth-Marnat and M. Jeffs, "Personality Factors from the Five-factor Model of Personality that Predict Dissociative Tendencies in a Clinical Population," *Personality and Individual Differences* 32 (2002): 969–976; and G. Groth-Marnat, R. I. Roberts, and L. E. Beutler, "Client Characteristics and Psychotherapy: Perspectives, Support, Interactions, and Implications for Training," *Australian Psychologist* 36 (2001): 115–121. Also see Alan R. Harkness and Scott O. Lilienfeld, "Individual Differences for Treatment Planning Personality Traits," *Psychological Assessment* 9 (1997): 349–360.

10. Groth-Marnat, Roberts, and Beutler, "Client Characteristics and Psychotherapy"; and Harkness and Lilienfeld, "Individual Differences."

11. PsychInfo is a Web site that provides access to a large number of peer-reviewed research journals, sponsored by the American Psychological Society.

12. For research on "affectionate positivity" measure, see especially G. S. Pettit and J. E. Bates, "Family Interaction Patterns and Children's Behavior Problems from Infancy to Four Years," *Developmental Psychology* 25 (1989): 413–420; and M. D. Ainsworth, "Attachments beyond Infancy," *American Psychologist* 44 (1989): 709–716, for a description of the experience of the "joy and pleasure" that characterizes affectional bonding and that is measurable upon reunion with a parent or other object of attachment.

13. The exercise described in chapter 5 is an example of an exposure procedure.

14. Foa and Kozak, "Emotional Processing of Fear," 20–35.

15. Edna B. Foa and S.A.M. Rauch, "Cognitive Changes during Prolonged Exposure versus Prolonged Exposure Plus Cognitive Restructuring in Female Assault Survivors with Post-traumatic Stress Disorder," *Journal of Consulting and Clinical Psychology* 72 (2004): 879–884. Foa, Hembree, Cahill, et al., "Randomized Trial of Prolonged Exposure."

16. See the self-rating scale used in exposure exercises described in chapter 5.

17. See, for example, Lambert, "Psychotherapy Outcome Research."

18. J. D. Safran and J. C. Muran, "Has the Concept of the Therapeutic Alliance Outlived Its Usefulness? *Psychotherapy: Theory, Research, Practice, Training* 43 (2006): 286–291.

19. For a very good review of how to work with a damaged therapeutic alliance, see Adam O. Horvath and L. Luborsky, "The Role of the Therapeutic Alliance in Psychotherapy," *Journal of Consulting and Clinical Psychology* 61 (1993): 561–573; and J. D. Safran and J. C. Muran, "The Resolution of Ruptures in the Therapeutic Alliance," *Journal of Consulting and Clinical Psychology* 64 (1996): 447–458.

20. The basis for this tendency is well articulated and explored in Weiss and Sampson's control-mastery theory. See, for example, A. Rappoport, "The Structure of Psychotherapy: Control-mastery Theory's Diagnostic Plan Formulation," *Psychotherapy: Theory, Research, Practice, Training* 33 (1996): 1–10.

21. Especially recommended is Marsha Linehan, *Skills Training Manual for Treating Borderline Personality Disorder* (New York: Guilford Press, 1993), from page 84 onward.

22. In addition to the above, good examples of the works of these pioneers include: Kernberg, *Borderline Conditions and Pathological Narcissism*; Christine

Padesky and Dennis Greenberger, *Mind over Mood* (New York: Guilford Press, 1995); and Young, *Cognitive Therapy for Personality Disorders.*

23. See Elizabeth A. Phelps, "Emotion and Cognition: Insights from Studies of the Human Amygdala," in *Annual Review of Psychology* 57, edited by Susan T. Fiske, Alan E. Kazdin, and Daniel L. Schacter (Palo Alto: Annual Reviews, 2006): 27–53, for a review of relevant neurobiological research.

24. Ibid.

25. The discovery of antibiotics is also another obvious and very major factor.

26. Practices that we outlined in the last chapter.

Chapter 9: Making Unscripted Emotions Our Allies

1. Some readers will recognize that "wrathful" is a word used to describe deities of similar appearance in Vajrayana teachings. Our use of the term is much more limited than that which is found in Vajrayana Buddhism, and so the two are comparable only in superficial ways.

2. In examining situational emotions, including anger, we are encountering an aspect of our experience that is also, at least partially, mediated by the amygdala. As a result, situational emotions, like emotions that arise from amygdala scripts, also act fairly independently of the consciousness-producing parts of our brain.

3. Two good examples of books that address the stages of grieving are Peter McWilliams, Harold H. Bloomfield, and Melba Colgrove, *How To Survive the Loss of a Love* (Santa Monica, Calif.: Prelude Press, 1973); and Bruce Fisher, *Rebuilding When Your Relationship Ends* (Atascadero, Calif.: Impact, 2000).

4. See Goleman, *Emotional Intelligence.*

5. See Thomas Geoff and Garth Fletcher, "Mind Reading Accuracy in Intimate Relationships: Assessing the Roles of the Relationship, the Target and the Judge," *Journal of Personality and Social Psychology* 85 (2003): 1079–1094; W. Ickes and J. A. Simpson, "Motivational Aspects of Empathic Accuracy," in *Emotion and Motivation,* edited by Marilynn B. Brewer and Miles Hewstone (Malden, Mass.: Blackwell, 2004), 225–246; and Goleman, *Emotional Intelligence.*

6. Vittorio Gallese, "Intentional Attunement: The Mirror Neuron System and Its Role in Interpersonal Relations," paper presented at the European Science Foundation Conference: "What Do Mirror Neurons Mean?" Paris, November 15, 2004; and Susan Hurley, "The Shared Circuits Model: How Control Mirroring and Simulation Can Enable Imitation and Mind Reading," paper presented at the European Science Foundation Conference: "What Do Mirror Neurons Mean?" Paris, February 15, 2005.

7. See, for example, Una McCluskey, Carol-Ann Hooper, and Liza Bingley-Miller, "Goal-corrected Empathetic Attunement: Developing and Rating the Concept within an Attachment Perspective," *Psychotherapy: Theory, Research, Practice, Training 1999 by the Division of Psychotherapy (29), American Psychological Association* 36 (1999): 80–90.

8. For a review of some of the relevant research on autism and empathy, see L. M. Oberman and V. S. Ramachandran, "The Simulating Social Mind: The Role

of the Mirror Neuron System and Simulation in the Social and Communicative Deficits of Autism Spectrum Disorders," *Psychological Bulletin* 133 (2007): 310–327.

Appendix B

1. For a review of some of the relevant research, see Amit Anand and Anantha Shekhar, "Brain Imaging Studies in Mood and Anxiety Disorders," *Special Emphasis on the Amygdala, Annals of the New York Academy of Sciences* 985 (2003): 370–388.

2. For neurobiological research on anxiety-related disorders, see M. J. Green and G. S. Malhi, "Neural Mechanisms of the Cognitive Control of Emotion," *Acta Neuropsychiatrica* 18 (2006): 144–153; De Raedt, "Does Neuroscience Hold Promise?"; and K. A. Corcoran and G. J. Quirk, "Recalling Safety: Cooperative Functions of the Ventromedial Prefrontal Cortex and the Hippocampus in Extinction," *CNS Spectrums* 12 (2007): 200–206.

3. For neurobiological research on depressive disorders, see W. Ramel, P. R. Goldin, L. T. Eyler, G. G. Brown, I. H. Gotlib, and J. R. McQuaid, "Amygdala Reactivity and Mood-congruent Memory in Individuals at Risk for Depressive Relapse," *Biological Psychiatry* 61 (2007): 231–239; U. Dannlowski, P. Ohrmann, J. Bauer, et al., "Amygdala Reactivity to Masked Negative Faces Is Associated with Automatic Judgmental Bias in Major Depression: A 3 T fMRI Study," *Journal of Psychiatry and Neuroscience* 32 (2007): 423–429; De Raedt, "Does Neuroscience Hold Promise?"; and W. C. Drevets, "Neuroimaging and Neuropathological Studies of Depression: Implications for the Cognitive-emotional Features of Mood Disorders," *Current Opinion Neurobiology* 11 (2001): 240–249.

4. For neurobiological research on borderline personality disorder, see Beblo et al., "Functional MRI Correlates."

5. For neurobiological research on addictions, see Gregory J. Quirk and Donald R. Gehlert, "Inhibition of the Amygdala: Key to Pathological States?" *Special Emphasis on the Amygdala, Annals of the New York Academy of Sciences* 985 (2003): 263–272.

6. Examples of researchers and their relevant research reviews include LeDoux, *The Emotional Brain*; Carlsson et al., "Fear and the Amygdala"; Dolan and Vuillemier, "Amygdala Automaticity in Emotional Processing"; and Phan et al., "Functional Neuroanatomy of Emotion."

7. See, for example, Carlsson et al., "Fear and the Amygdala."

8. For a review of this research, see Adolphs, Denburg, and Tranel, "Amygdala's Role in Long-term Declarative Memory," 983–992; Denburg et al., "Preserved Emotional Memory"; and Dolcos, LaBar, and Cabeza, "Remembering One Year Later."

9. See for example, Adolphs, Denburg, and Tranel, "Amygdala's Role in Long-term Declarative Memory."

10. Dolcos, LaBar, and Cabeza, "Remembering One Year Later."

11. Denburg et al., "Preserved Emotional Memory."

12. See, for example, Ann E. McIntyre, Benno Roozendaal, and L. McGaugh, "Role of the Basolateral Amygdala in Memory Consolidation," *Special Emphasis on the Amygdala, Annals of the New York Academy of Sciences* 985 (2003): 273–293.

13. Phelps, "Emotion and Cognition."

14. Ibid.

15. A. Öhman and S. Mineka, "Fears, Phobias, and Preparedness: Toward an Evolved Module of Fear and Fear Learning," *Psychological Review* 108 (2001): 483–522.

16. See, for example, the summaries provided by Green and Malhi, "Neural Mechanisms"; and Buchanan, "Retrieval of Emotional Memories."

17. De Raedt, "Does Neuroscience Hold Promise?"

18. Green and Malhi, "Neural Mechanisms," 144.

19. Davidson, Jackson, and Kalin, "Emotion, Plasticity, Context, and Regulation," 891.

20. Öhman and Mineka, "Fears, Phobias, and Preparedness," 485.

21. Phelps, "Emotion and Cognition," 41.

22. See De Raedt, "Does Neuroscience Hold Promise?"; Phan et al., "Functional Neuroanatomy of Emotion"; and Quirk and Gehlert, "Inhibition of the Amygdala."

23. See Davidson, Jackson, and Kalin, "Emotion, Plasticity, Context, and Regulation"; Ramel et al., "Amygdala Reactivity and Mood-congruent Memory"; Carlsson et al., "Fear and the Amygdala."

24. See Lieberman et al., "Putting Feelings into Words"; Creswell, Way, Eisenberger, and Lieberman, "Neural Correlates of Dispositional Mindfulness," *Psychosomatic Medicine* 69 (2007): 560–565; Brown and Ryan, "Benefits of Being Present."

25. See, for example, A. D. Craig, "Human Feelings: Why Are Some More Aware than Others?" *Trends in Cognitive Sciences* 8 (2004): 239–241; Critchley et al., "Neural Systems Supporting Interoceptive Awareness"; and Pollatos, Gramann, and Schandry, "Neural Systems Connecting Interoceptive Awareness and Feelings."

26. See B. K. Hölzel, U. Ott, T. Gard et al., "Investigation of Mindfulness Meditation Practitioners with Voxel-based Morphometry," *Social Cognitive and Affective Neuroscience* 3 (2008): 55–61;. Brown and Ryan, "Benefits of Being Present."

27. Phelps, "Human Emotion and Memory"; and Phelps, "Emotion and Cognition"; De Raedt, "Does Neuroscience Hold Promise?"; and Quirk and Gehlert, "Inhibition of the Amygdala."

28. See, for example, Davidson, Jackson, and Kalin, "Emotion, Plasticity, Context, and Regulation," 893.

29. See, for example Adolphs, Denburg, and Tranel, "Amygdala's Role in Long-term Declarative Memory."

30. For "implicit memory," see for example, D. B. Fenker, B. H. Schott, A. Richardson-Klavehn, H. Heinze, and E. Düzel, "Recapitulating Emotional Context: Activity of Amygdala, Hippocampus and Fusiform Cortex during Recollection and Familiarity," *European Journal of Neuroscience* 21 (2005): 1993–1999; for "gist," see, for example, Adolphs, Denburg, and Tranel, "Amygdala's Role in Long-Term Declarative Memory." For a good summary of this research, see Jacobs and Nadel, "Neurobiology of Reconstructed Memories," 1110–1134.

31. Jacobs and Nadel, "Neurobiology of Reconstructed Memories," 1127.

32. See, for example, De Raedt, "Does Neuroscience Hold Promise?" See Buchanan, "Retrieval of Emotional Memories," for a review of this research.

33. See Buchanan, "Retrieval of Emotional Memories," for a review of relevant research.

34. See De Raedt, "Does Neuroscience Hold Promise?" for a review of some of this research.

35. Phelps, "Human Emotion and Memory," 198.

36. Buchanan, "Retrieval of Emotional Memories," 776.

37. Davidson, Jackson, and Kalin, "Emotion, Plasticity, Context, and Regulation," 903.

38. De Raedt, "Does Neuroscience Hold Promise?"

39. See Green and Malhi, "Neural Mechanisms."

40. Phelps, "Emotion and Cognition." 45.

41. See Critchley et al., "Neural Systems Supporting Interoceptive Awareness"; Pollatos, Gramann, and Schandry, "Neural Systems Connecting Interoceptive Awareness and Feelings"; and Craig, "Human Feelings."

42. See, for example, Davidson, Jackson, and Kalin, "Emotion, Plasticity, Context, and Regulation."

43. See, for example, Creswell et al., "Neural Correlates of Dispositional Mindfulness."

INDEX

Age of Reason, 57
amygdala, neuroscience of, xv, 2–4, 49, 161, 169–176
amygdala scripts: activation of, 8–9, 14, 90; advantages of, as therapeutic tools, 123–147; belief as one component of, 22–23, 30–31, 51, 100–122; characteristics of, 40–41; childhood and, 35, 177–179; defensiveness and, 5, 19–23, 26, 31, 42, 64, 82, 132, 155, 173; definition of, 6–8; disappearance of, after following Three-Step Practice, 144–145; emotion as one component of, 19–20, 23–24, 28, 54–75, 83–84, 114, 137, 147, 171; evolutionary theory and, 7–8, 35–40, 45–47, 76–77, 80 (*see also* amygdala scripts: defensiveness and); exposure therapy and, 137–138; identification of, 64–65; image as one component of, 11–12, 26–28, 76–99, 114, 137–138, 172–173; insight therapy and, xi, xii, xiv, xv, 13–14, 42, 84–85; mastery of, as a goal, 14, 18, 40, 46–47, 63, 96–98, 113, 116–118, 135, 141, 145–146, 156, 171, 173–175; metascripts and, 119–120, 140–141, 143; neuroscience of, xi–xii, 3–5, 10–11, 33, 35–36, 38, 42–43, 78–79, 92–93, 120, 145, 169–176; "neurotic complexes" and, 7; understanding of, in a pharmaceutical age, 15, 35; unscripted emotions and, 147, 152, 154, 156–158, 160. *See also* emotion memories
anger (scripted response), 7, 20–21, 24, 46–47, 49, 92, 101, 119

anger (unscripted response), 149–155
anxiety (scripted response), 7, 10, 14, 17–18, 24, 30, 35, 44, 60–61, 64, 83, 87, 90–92, 97, 118–119, 129, 142, 169, 171
Aristotle, 56
Atlantic Monthly, 5
Augustine (saint), 57
autism, 162

Beck, Aaron, xi, 103
behaviorism, xi, xii, 8–9. *See also* cognitive therapy
belief component (of amygdala script), 22–23, 30–31, 51, 100–122
bipolar disorder, 15, 43, 128
Bordin, Edward, 125
Brown, Kirk Warren, 172
Buchanan, Tony, 170
Buddhism, 49, 59–61, 64, 106, 121

Cabeza, Roberto, 170
Cartesian dualism. *See* mind/body dilemma
categorical diagnosis, 129. *See also* psychopathology
cerebral cortex: conscious awareness and, 3–4, 6, 37–38, 54–55, 59, 80–83; in dealing with unscripted emotions, 153, 157–159 (*see also* unscripted emotions); in mastering amygdala scripts, 8, 11–13, 25, 41, 72–74, 97–98, 102, 105, 137, 153, 172, 174–175; neuroscience of, xv, 170
client-centered psychotherapy, xii

ABOUT THE AUTHOR

Timothy B. Stokes, PhD, has been doing psychotherapy and providing clinical consultations to other therapists for more than thirty years. He is also the owner and clinical director of Corporate Psychological Services, a company that has provided psychological services to thousands of people since 1985. His interest in psychological writing was prompted in 1990 when he began a five-year stint as the editor-in-chief of the *Naropa Journal of Contemplative Psychotherapy*. His thirty-five years as a Vajryanna Buddhist meditator, along with the good fortune of having learned from many very accomplished teachers have also served to shape Dr. Stokes's understanding of the human mind.